IMMUNE

Dr Servaas Bingé is a general practitioner and sports physician. From his work as chief physician of Lotto-Soudal, the oldest professional cycling team in the world, he developed the concept of 'healthitude', combining our health, attitudes, and daily choices. He is a sought-after speaker on topics of health and is the author of four books.

Anouk Brackenier is a certified health coach with a PhD in biotechnology. She is passionate about the impact of lifestyle on health, and has created recipes for three books in partnership with Servaas.

David Shaw works as a journalist for Germany's international broadcaster, Deutsche Welle, as well as translating from several languages, including German, Dutch, Russian, and French. He lives in Berlin.

Dr Servaas Bingé

Translated by David Shaw

IMMUNE

stay healthy and take good care of your immune system

SCRIBE

Melbourne • London

Scribe Publications
2 John Street, Clerkenwell, London, WC1N 2ES, United Kingdom
18–20 Edward St, Brunswick, Victoria 3056, Australia
3754 Pleasant Ave, Suite 100, Minneapolis, Minnesota 55409, USA

First published in Dutch as *Immuun: Blijf gezond, bewaak je weerstand*
by Borgerhoff & Lamberigts, Belgium in 2020

Published in English by Scribe 2022

The advice in this book is not intended to replace the services of trained health pro-
fessionals or be a substitute for medical advice. You are advised to consult with your
health care professional with regard to matters relating to your health, and in particular
regarding matters that may require diagnosis or medical attention. Names have been
changed, identifying details and backgrounds have been altered, and stories combined out
of respect for the privacy of individuals and couples.

Typeset in Portrait by the publishers.

Printed and bound in the UK by CPI Group (UK) Ltd, Croydon CR0 4YY

Scribe is committed to the sustainable use of natural resources and the use of paper
products made responsibly from those resources.

978 1 913348 97 7 (UK edition)
978 1 950354 90 0 (US edition)
978 1 922310 98 9 (Australian edition)
978 1 922586 33 9 (ebook)

Catalogue records for this book are available from the National Library of Australia and
the British Library.

scribepublications.co.uk
scribepublications.com
scribepublications.com.au

CONTENTS

'May the force be with you'

Master Yoda

INTRODUCTION

These Here Are Crazy Times

2020. From one week to the next, everyone was suddenly bandying about words like 'lockdown', 'face mask', 'travel bubble', 'social distancing', and 'quarantine' as if they'd been using them all their lives. Suddenly, 11-year-old kids were telling us adults off for not disinfecting our hands after entering an enclosed space. 'Restrictions', 'tier one/two/ three', and 'red/orange/green zone' became much-googled terms. The coronavirus had us waving at each other over our garden hedges, mime-hugging at two metres' distance, working from home, and learning over Zoom. Not since World War II have children spent so many months away from school. And not since the 1980s have our roads seen so little traffic. One bat urinated in the wrong place and in no time at all the entire world was in the grip of COVID-19.

1918. The so-called Spanish flu swept the planet. That pandemic claimed millions of lives. Face masks and isolation were also then promoted as effective ways to stop the bug spreading.

Has so little changed, basically, in over a hundred years? Are we as defenceless now as we were then? Of course not. Although COVID-19 has spread like wildfire, is difficult to get to grips with, and still harbours many unknowns, we certainly know a great deal more than we did a hundred years ago.

In 1918, the profession of virologist was still unheard of. Scientists knew almost nothing about viruses, and we had no idea how they behave inside our bodies. Isolating patients was proven to be an effective measure not by science, but through experience. We'd known that since the Middle Ages, when the bubonic plague rampaged through Europe. By the 20th century, handwashing and airing rooms were seen as simply commonsense behaviour, and together with isolation they are as effective today as they were then. All these precautions had an influence on the spread of the disease. Even a hundred years ago, it was easily apparent that some people get sick while others don't, or develop far less severe symptoms when they do. It wasn't until later that we began to discover that there was such a thing as the immune system, and that having a strong immune system was important for our continued good health. I'm going to be talking a lot about that amazing system in this book.

So Everyone's a Virologist Now?

I'll be bombarding you with phagocytes, lymph nodes, bloodstreams, front lines, attackers, and defenders. Getting to know this tiny world inside you will help you better understand how the dangers from the big world outside can affect your body. So, will you be able to declare yourself a self-taught virologist after reading this book? Absolutely not! My aim is simply to provide a little insight into this complex world. I want to provide you with a clear explanation of your immune system: what it is, how it works, what it does, and how you can keep it functioning at its best. I want you to understand it, through both common sense and science. The latter may lag behind a little — but I don't mean that as a reproach. Let's combine common sense and science, in a language we can all understand.

What Is the Immune System, Anyway?

The immune system is one of the body's biological systems, like the digestive and nervous systems, for example. You can't consciously control it, just as you can't control the workings of your heart or kidneys. It's just there, doing its job.

The immune system is our body's fortress. It's a network of cells, tissues, and organs that protects us against foreign invaders and any other threat to our health. We all know that such a threat can come from an airborne virus, but it could just as easily come from a mosquito bite. Without the defences provided by our immune system, an insect bite like that could be fatal.

The lead actors in the immune system are the white blood cells, which are produced in the bone marrow of our spinal cord and circulate through our bloodstream. White blood cells are our security guards, always on patrol, keeping an eye on who (or what) can come in and who (or what) can't.

There are two main kinds of white blood cells. The first are phagocytes (from the Greek *phagein*, 'to eat', with the suffix *cyte*, meaning 'cell'), which devour anything that doesn't belong in our bloodstream. A bit like so many little chomping Pac-Mans. After gobbling up an invader, they process its information, which they pass on to the second

3

main kind of white blood cell, the lymphocytes. Depending on the information they receive, these lymphocytes transform themselves into one of two subtypes.

These lymphocyte subtypes are the T cells and B cells. If our immune system is a fortress, the T cells are the knights protecting it, ready to attack the enemy at a moment's notice. B cells are the factories that produce antibodies to protect us — permanently or temporarily — against the enemy. To use a concrete example: if your doctor tells you she's going to take a blood sample to test whether you have antibodies against the coronavirus, she's checking how hard your B cells have been working.

Naturally, the above summary of how the immune system works is far too brief and requires further explanation. It's important to understand the underlying mechanisms if you want to strengthen them. We are currently bombarded with tips and tricks on how to 'boost' our immunity, which is something you might say we are in sore need of — living as we are in the time of corona! But, to be honest, I'm no great believer in such short-term 'boosts'. In this book, I aim to provide a detailed but also plain and easily understandable explanation of how we can make sure our immune system stays as healthy as possible. Because, although I don't believe in 'boosting' the immune system, I believe there is certainly lots we can do to influence it.

First, we can take to heart the measures imposed on us — now, during the coronavirus pandemic, but also during the flu season, and other times, too. Washing our hands regularly, maintaining social distancing, staying at home if we feel sick — you know the drill! When it comes down to it, the fewer invaders you give the opportunity to get inside your body, the more likely you are to remain healthy.

Second, blood tests can identify what your immune system needs, what it's lacking in order to work at its best, and what elements it's deficient in. That leads us seamlessly to point three, and anyone who knows me will already have guessed what hobbyhorse I'm about to conjure out of my hat.

By optimising our lifestyle and tackling the classic issues of diet, sleep, exercise, and stress, we can really improve and increase our immunity!

Come with me and I'll tell you all about the great, but also microscopically small, threats that come from both without and within. We're about to enter the microscopic world of viruses, bacteria, vaccinations, and immunity, and our story begins around the middle of the 18th century.

Vaccination, Hygiene, and Medicine

Smallpox is an infectious disease of the respiratory tract that affects our organs via the bloodstream and can cause high fever and even death. In the 1700s, it was one of the most feared diseases. Early attempts to control smallpox involved the process of variolation, or inoculation, in which a less-severe version of the disease was introduced into the skin, after which the patient was kept in isolation until their illness had run its course. The risk of death was lower than with the severe form of the disease, but not zero.

Then, one day, a doctor by the name of Edward Jenner discovered that milkmaids who had been infected with the less-virulent disease cowpox appeared subsequently to be immune to the human variant, smallpox. To test his hypothesis, he used his neighbour's son as a guinea pig and injected him with a small amount of the fluid taken from a cowpox pustule. Obviously, medical ethics commissions were not a thing at the time ...

The boy developed cowpox, became moderately ill, as expected, and, after recovering, was given an injection of human pox. Jenner had performed the first documented vaccination. What happened next was ... nothing. The boy was immune. And, as is so often the case with great medical advances, this innovation was not accepted by medical science without a struggle. Jenner was maligned and silenced, and it was not until 50 years later that the traumatic technique of variolation was abandoned and replaced by vaccination. The world could finally be immunised against smallpox.

Alongside vaccination, more attention was gradually paid to hygiene. Nowadays, we take washing our hands for granted, but that wasn't always the case. In the mid-19th century, many women died of puerperal fever after giving birth. The Hungarian obstetrician Ignaz Semmelweis began to study this phenomenon and soon established a link between the incidence of the disease and the unhygienic habits of doctors. After requiring all physicians to wash and disinfect both their hands and their equipment, he observed a dramatic fall in the mortality rate. The medical world baulked at this, and Semmelweis was also vilified and called insane. He eventually died impoverished in a mental institution. It wasn't until decades later that scientific research into the connection between puerperal fever and the hygiene of doctors and their equipment proved Semmelweis right. Subsequently, it was also discovered that sterilisation could kill bacteria, and recognition gradually grew of the importance of our own, individual contribution to both personal and public hygiene.

A third great leap of progress was the development of medicines and chemical substances to combat bacteria and viruses. Paul Ehrlich accidentally came across a medicine to treat syphilis, and a short time later it was found that the drug was most effective when administered intravenously. Other medicines were also developed, such as antibiotics — it was Alexander Fleming who discovered penicillin, again by accident, when he noticed that a fungus was highly effective at combating bacteria.

But why am I telling you all this? These are important milestones in the history of medicine: they have helped us understand how the immune system works and contributed to health protection. Not much more than a hundred years ago, we were still dying of 'common' infectious diseases. Now we expect potentially contaminating objects to be disinfected and sterilised. Now we can be cured with a simple pill or injection.

That's also the reason why the whole world is now looking on in shock as the COVID-19 tsunami crashes over us. We thought dying of infectious diseases was a thing of the past, but it turns out that we don't have a ready answer for everything, after all. Over the past few decades, we have seen a few such diseases: just think of bird flu, swine flu, and SARS.

COVID-19

COVID-19 is not a unique virus. Coronaviruses are not new: they are common and come in different sorts. They are mostly harmless, causing nothing more serious than the common cold. For some reason, this specific virus mutated in such a way that it's able to cause an acute respiratory syndrome — something we have already seen happen twice recently: the mysterious SARS (severe acute respiratory syndrome) virus suddenly appeared in Asia in 2002; and in 2012 it was the MERS (Middle East respiratory syndrome) virus, which circulated mainly in that part of the world (and has still not completely disappeared). Like COVID-19, both are variants of the coronavirus that passed from animals to humans. In fact, the scientific names of the SARS, MERS, and COVID-19 viruses are SARS-CoV-1, MERS-CoV, and SARS-CoV-2.

How come these viruses are so infectious? They can be passed on in two different ways. One is known as faecal–oral transmission. For example, if you use a toilet after someone who is infected, don't wash your hands, and then touch your nose or mouth. However, the more common method of transmission is via airborne particles, ranging from larger saliva droplets to smaller aerosols. Whenever we speak, we can easily spread droplets over a distance of a metre and a half. That's why wearing a mask is so important in situations where maintaining a social distance of at least a metre and a half is not possible.

How does an infection happen? The particles containing the virus float through the air and come into contact with our mucous membranes, where they enter our cells. Mucous membranes are found under your eyelids, inside your nose, lips, and mouth, and deeper down in the respiratory tract. That is this particular virus's first point of attack on the fortress of our immune system.

But before we delve deeper into the subject of our immune system, it is time, gentle reader, for a crash course in epidemiology! When a new virus like COVID-19 emerges, it's important to understand it as quickly as possible. When does the disease occur? How is it transmitted from person to person? What are its causes? How does the resulting disease progress? These are questions for the science of epidemiology.

Stopping the Spread

A disease becomes an epidemic when it spreads rapidly to many people, or to a group of people in the same area. An epidemic of gastroenteritis at a scout camp can be a relatively small epidemic, but, every year in the cold months, we also have a flu epidemic that affects a larger group of people. When an epidemic becomes so widespread that it spans the entire world, we call it a pandemic.

Doctors who encounter infectious diseases have a duty to report them. That information is saved in databases and statistical records, and is expressed in terms of incidence and prevalence. The prevalence of a disease is the number of people infected with it at a given point in time, expressed as a proportion of the entire population. The incidence is the number of confirmed *new* infections in a given group of people during a defined period of time.

So, in the case of COVID-19, the incidence is the figure published every day giving the number of new cases, and the prevalence is the percentage of people who have COVID-19 at that specific moment.

Another interesting concept is the famous R number we hear so much about when each new variant of COVID-19 starts to spread. The R number is a measure of infectivity, and for the original strains of COVID-19 it lies somewhere between two and three. That means every infected person will infect three more. Those three will become sick and go on to infect another three people each. When that happens, the number of infections rises rapidly. The seasonal flu, by comparison, has an R number of 1.3 and so spreads far more slowly and gradually.

Hospitals are a key domain of epidemiology. Patients sometimes catch diseases that have nothing to do with the conditions they've come in for, including respiratory or mouth infections, or the dramatic flesh-eating bacteria. Of course, the most important cause of those infections is that the immune systems of people who are already sick isn't functioning fully, while other patients have to take medication that suppresses their immune response. In addition, microbes often develop resistance to certain drugs — such as antibiotics. By carrying out epidemiological research on patients, we can learn to reduce the risk of infection to as low a level as possible.

One strategy is to keep infected patients in isolation. The hygiene rules for staff members are may also be strict, and workers are tested for infections regularly. Ideally, everyone in a hospital should behave as if everyone else were carrying an infectious disease. The approach taken with COVID-19 is a good example of that epidemiological policy. Where possible, each patient is treated or tested on a separate ward, and the doctors and nurses wear PPE — personal protective equipment: suits, masks, gloves. The same measures have been used for years when treating people who have no functioning immune system.

Stopping the Spread, Part II

But, beyond the hospital environment, how can we protect ourselves from illness? To use a rather scientific-sounding term: with prophylaxis on a social scale.

Sorry, what?

We have to prevent people being able to get sick in the first place. One way we can do this is by making sure the reservoir of available pathogens remains small. A pathogen is something that causes disease: a virus, bacterium, or fungus, for instance. The fewer pathogens there are, the smaller their chance of spreading is. A clean water supply is extremely important because diseases like typhoid fever, cholera, and gastroenteritis are transmitted via water. We should rinse vegetables and fruit thoroughly to reduce the chance of getting food poisoning from *E. coli, Listeria,* and salmonella. We protect ourselves against insects, which can spread malaria, and do all we can to stop them reproducing. And then, of course, there's personal hygiene, including things like washing your hands — and other parts of your body from time to time, if you don't mind!

Another way we can prevent disease is by eradicating the disease in question. The most effective way to achieve this is still by immunising the entire population. This can be done passively, by administering antibodies that give temporary protection against a specific disease or (with greater risk and no guarantee of success) by allowing people to catch the disease so that the body builds up an immunity to it. More importantly, prevention can also be tackled actively, by using vaccination to immunise a population against a certain illness.

By the way, do you know where the word *vaccine* comes from? The answer takes us back to Mr Jenner, whose story I told earlier. He injected another human being with fluid taken from a cowpox blister. And the Latin word for cow is ... *vacca!*

There are several different types of vaccination:

- In the first, the patient is injected with an attenuated (weakened) version of the pathogen, to stimulate a mild infection in the body. No serious disease occurs, but the patient receives a robust and prolonged immunity. Vaccines against tuberculosis, polio, measles, mumps, and rubella are examples of this.

- The second type of vaccination is with inactivated vaccines, in which only a certain part of a virus or bacterium is introduced into the person's body. The immune system detects an invader and starts producing antibodies, without the patient having to go through the illness. Rabies, typhoid fever, and seasonal flu are prevented using inactivated vaccines.

- The third type of vaccination uses genetically produced vaccines to provoke a reaction in the body. No material from the pathogen is used. The hepatitis-B vaccine works this way. Anyone planning to travel abroad or work in the health sector must be vaccinated against hepatitis B. A series of three vaccines or boosters provides long-term immunity from the disease.

- The fourth type makes use of toxoids, which are toxins found in pathogens. Such vaccines use only the toxoids, not the pathogen itself, to trigger an immune response. The tetanus vaccine is a good example of this kind of vaccination.

Anti-vaxxers

Are we finally going to talk about the immune system now, Doctor?

Well, there's one more important issue I have to address before we can really get going. I can't write a book about the immune system without putting the vaccination debate on the table. That would be like a bicycle without its wheels, or Pippi Longstocking without her braids ...

It's a difficult issue, admittedly.

We're expected to have an opinion on everything these days. What do you think about food product X? Do you think Y is equally good for everybody? And so on ... I have made it clear on more than one occasion that that's not how I live my life and it's not the way I see the world. Why does everything always have to be so polarised? After all, the truth is never just black or white. What follows is not an opinion, but rather a series of facts. Based on my experience and my knowledge. Everyone is at liberty to choose for themselves what they do; who am I to impose anything on anyone? All I can do is to advise you as honestly as possible.

The fact that, in the course of the history of medicine, large-scale vaccination campaigns are proven to have eradicated certain diseases from the world is not something we need to argue over. Yet it's also clear that concerns remain, among both supporters and opponents.

Some parents are afraid of doing something 'wrong'. They believe vaccinations can harm their child, after reading various posts on the internet, for example. On the other side are the epidemiologists, doctors, and health experts, who have seen vaccinations save lives and believe that the advantages outweigh the risks. For certain diseases, they support the strategy of creating herd immunity by vaccinating people, and thus eradicating those diseases from the population. However, eradicating a disease from a population does not mean that the disease no longer exists.

Leaving children unvaccinated has consequences. As a parent, you can also put your children at risk by not vaccinating them, as it exposes them to a perfectly avoidable danger. What you are actually doing then, as a parent, is counting on all other parents to protect *your* children indirectly by having *their* children vaccinated and thus maintaining herd immunity. The greater the number of people who refuse to vaccinate their children, the greater the danger that unvaccinated children will get sick and become spreaders themselves. That's why the viral disease that causes measles is now on the rise again, despite the fact that there's a vaccine against it. Just imagine if you were the parent of the one or two children in a thousand who do not survive a bout of measles.

That's why I believe it's important to follow the program of basic vaccinations for young children closely. The fact that some diseases no longer occur doesn't mean that those diseases no longer exist per se; it simply means that herd immunity has been achieved and the vaccination works. Immunity will only last as long as we continue vaccinating.

It's the Dose That Makes the Poison

Those comments are about parents and children now, but the anti-vaccination movement isn't only a 21st-century phenomenon. On the contrary. Anti-vaxxers existed even before the first vaccine was invented. Edward Jenner — there he is again — experimented with cowpox around the year 1796, but, even before that, there was the technique of variolation, which I mentioned briefly: it's the application of fluid from smallpox pustules to the skin, to trigger the disease and subsequent immunity. Even back then, a group of 'anti-inoculators' protested against the practice, claiming it spread fear and paranoia among the population. There were more protests when the first smallpox vaccine was developed, although vaccination is a much safer technique than unhygienic variolation. The anti-vaccination movement made claims of medical despotism, warned of the toxic chemicals used in vaccines, and promoted alternative remedies, such as homeopathy. It wasn't until the 20th century — from 1943, to be precise — that other vaccines started to be developed (leaving aside Pasteur and his discovery of a treatment for rabies) and large-scale protest movements began. The anti-vaccination movement that exists today is still convinced that big pharma wants to make people sick on purpose by lacing vaccines with poison and toxins in order to make us buy more of its drug products.

On the other hand, the anti-vax movement could also be seen as big business, since it promotes the sale of vitamins and supplements under the guise of immunity boosters, which often simply don't work.

Vaccines do indeed contain certain substances, such as preservatives, which could be toxic in very high doses. You might be wondering why. The compounds in the vaccine need to be kept stable. It might be that such a substance is currently the only one able to preserve the element that provides immunity effectively. The following list describes the contents of a vaccine more specifically:

- the antigens that cause our immune system to switch to alarm mode and start producing antibodies
- fluid — sterile water or saline solution — to keep the other substances together and make the vaccine ingestible or injectable
- adjuvants, which are additives that help the body develop a stronger immune response
- preservatives or stabilisers, which protect the vaccine against heat, moisture, or light, and without which the vaccine would not be suitable for the market.

The mercury contained in vaccines is not the same as the mercury that can cause brain damage if you ingest too much of it from contaminated fish. The mercury in vaccines is in the form of ethylmercury, in which it is bonded to an alcohol group such as methanol or ethanol. In vaccines, it serves to prevent contamination. Some vaccines also contain aluminium. But just notice the dose! Aluminium is a substance which can be processed by the body quickly in small amounts. There's more aluminium in breastmilk than in those vaccines. Formaldehyde is another preservative: it's toxic and can cause diseases, such as asthma. But formaldehyde occurs in nature, and so we are constantly breathing it in. The dose in vaccines is so very tiny that it really is harmless.

It's impossible to prove a negative. Can anyone prove that a child doesn't have a feared condition because they didn't get a shot? No. Such arguments will get us nowhere fast, I'm afraid, but it is in our nature as human beings.

Better, then, to base our discussion on what can be proven. On actual facts. On the truth.

- Vaccines deliver on what they promise: they prevent disease. Full stop. There is a good reason why rubella and polio have disappeared. These diseases are now virtually non-existent because the population has been vaccinated against them and those vaccines work.
- Vaccines protect more than just you. Vaccination helps achieve herd immunity, thus protecting others as well. This is really something we all have to do together!
- The consideration here is not about whether to get vaccinated against a simple sniffle. The diseases against which vaccines have been developed are serious illnesses that in the past usually meant hospitalisation and/or death.
- The anti-vaccination movement may seem big, but we still have more children who are vaccinated than not. Anti-vaxxers benefit from the fact that so many people do get vaccinated, as they, too, enjoy protection from the herd immunity achieved by others.
- Vaccines have undergone extremely stringent testing for their safety. Vaccination law is part of the legislation

covering medications, so vaccines are subject to far more testing than treatments promoted as alternatives. Indeed, some of those supplements may well be more unsafe than the vaccines they claim to replace.

At the moment, we find ourselves in a very special situation. For the first time in many years, vaccines are being launched on the market that haven't previously undergone years of research. The current situation makes this a necessity; we have no choice if we want to put a stop to COVID-19. That doesn't mean the vaccines haven't undergone rigorous scrutiny.

Before a vaccine is launched, all its ingredients are individually identified and tested to ascertain in what amounts they become toxic. Only after that research is complete is the vaccine tested on cell cultures. This helps predict how the body might react to it, and its make-up can be adapted if necessary. Only then is it tested on animals. Many vaccines fail to make it to this stage, as they turn out to be too harmful after all. Next, clinical studies are carried out, in a minimum of three phases. First, the vaccine is tested on small groups of subjects. Then it's tested on a larger group, which is monitored very closely; this is also when researchers examine in what form it's best administered. The third phase involves testing on a much larger scale and gathering lots of data. That information is collated and passed on to the relevant national regulatory authority, which either approves the vaccine or not. Use of the vaccine is then monitored over the next few years by a controlling body, and doctors are asked to report any possible side effects.

Each COVID-19 vaccine passes through all those stages, only much more quickly. None of the phases are skipped, as that would be unethical. But the procedure has been streamlined to involve less paperwork, so we can get down to the job of vaccinating the world.

How Do We Get Sick?

None of us is alone in this world. We share our planet with many other living things. In science, we often speak in terms of host and guest. If the two get on well together, we call it symbiosis — different organisms living together. This relationship can take various forms:

- In commensalism, one organism benefits from the relationship while the other is unaffected. It receives neither benefit nor harm from the situation. An example of this is arctic foxes following polar bears around, to eat what the larger animals leave behind of their prey.
- In mutualism, there's a win-win situation for the organisms involved. Both benefit from living together. For example, our gut is full of beneficial bacteria. We keep them alive and, in return, the little critters provide us with many useful services, as I will explain in great detail later in this book.
- In parasitism, one organism clearly benefits from the symbiotic relationship, while the other is harmed by it. If you have a tapeworm, for example, it will make you sick, or at least prevent you from getting enough nutrients from your food.

Sickness is caused by the pathogenic (illness-causing) properties of viruses, bacteria, or fungi, each of which has a certain level of virulence (intensity). Some bacteria aren't pathogenic and can't make us sick. Others are indeed pathogenic but have a low level of virulence and so don't pose much of a threat to our health. Pathogens with a high level of virulence can make us very ill indeed.

We all carry with us a reservoir of pathogenic germs. On our skin, in our respiratory tract, and in our gut. Our family pets are also full of pathogens — as are our herb gardens, for that matter. The world we live in is full of septic tanks! At any given moment, one of those germs can leave the reservoir it lives in and invade a human body. This is what we call disease transmission. Most germs make use of a specific portal of entry; each has its own way of invading us. In the case of COVID-19, the method of entry is via respiratory particles that find their way to our mucous membranes; in the case of sexually transmitted diseases, the portal of entry may be via the genitals. So, we have many different such portals: our nose, throat, eyes, and mouth; our digestive system, genitals, and urinary tract ...

In general, there are seven ways diseases can enter our bodies.

Pathogens can be transmitted from person to person, via larger saliva droplets but also via physical contact. This is known as horizontal transmission — via direct, person-to-person contact.

Another way is vertical transmission, which is also a kind of person-to-person transmission, but when the two people stand in a special relation to each other. It includes transmission from mother to child, during childbirth or later during breastfeeding. Mastitis is a skin infection common in new mothers and is often caused by the baby's saliva, which enters the mother's body via tiny wounds and causes inflammation of her breast or nipple.

I've already mentioned faecal–oral transmission. This could happen when someone doesn't wash their hands after going to the toilet, then takes a stroll hand in hand with their sweetheart. If said sweetheart then eats an apple, or rubs their eyes, transmission is a done deal. ('Their', 'his', and 'her' and are freely interchangeable in this context.)

Airborne transmission occurs when the pathogen is able to linger in the air for a long time and is simply inhaled without any human contact being involved. This is the way tuberculosis spreads — and a way that COVID-19 can spread, too.

Transmission via a vehicle is another possibility. The germs stick to the surface of something, such as a glass or a fork, or even a table, and enter the body that way.

Parenteral transmission occurs via blood-to-blood contact through an unnoticed lesion in the skin or mucous membrane: if you have a small skin wound on your hand and you shake hands with someone with a similar lesion on theirs, you can get infected.

Finally, we have transmission via living vectors, usually insects, such as the malaria mosquito, ticks, and tsetse flies.

Once the critters have contaminated us, they like to stick around. Adhesins are proteins found on the surface of some bacteria that make it easier for them to adhere to the portal of entry, thus increasing their virulence.

Disease develops when our defences are either overcome or bypassed. There are actually very few pathogens that can cause disease on the surface of our bodies, notwithstanding some bacteria that use our pores as a portal of entry. Most pathogens tend to penetrate deeper — in other words, they are more invasive. And they make use of various substances to help them do that.

Streptococci are bacteria found in the throat. They produce a certain enzyme that destroys hyaluronic acid — an important component of our connective tissue — enabling them to penetrate anywhere easily. The COVID-19 virus uses a different mode of infiltration. It binds to a certain receptor (ACE2) on the cells of the respiratory tract, tricking them into thinking they've received something useful as a gift. That's how the virus makes its way into our cells. Evil!

Some pathogens have the ability to bypass our defence mechanisms. The bacterium that causes tuberculosis can survive inside a phagocyte — you remember the Pac-Man-like white blood cells that gobble up any strangers in our bloodstream — and thus spread through the body.

Other pathogens, such as flesh-eating bacteria, use a certain protein on the surface of their cell walls to disguise themselves and so trick our bodies into thinking everything is normal. The disguise they put on is called M protein. By the time our body's reflex to remove that cloak of disguise has kicked in, the bacteria have already spread.

There are also organisms that mutate in order to bypass our defences. Viruses that cause the common cold, as well as coronaviruses, are very good at this. You can be immune to a virus, only for it to quickly don a different disguise and fool your body — and then it's got you. Small mutations can quickly mean a loss of immunity.

In this way, pathogens can spread superfast and reproduce throughout our body, where they begin to cause all kinds of havoc.

Now, let's return to our reservoirs, which are home to all sorts of different flora.

We have a very stable relationship with our resident flora. These are the microbial viruses, bacteria, and fungi that are present in our bodies for our entire lives, such as staphylococci. They live externally on our skin, but also internally in our mucous membranes or digestive system.

Our transient flora come and go. Sometimes they are on our body, sometimes not. Whenever we wash, they disappear. They aren't sufficiently adapted to become permanent residents. Many hospital bacteria are transient flora. Staphylococci, which are common in everyone, exist in a different form in hospitals, and that form is resistant to antibiotics. Insufficient handwashing in hospitals can spread the bacteria to those with weaker defences and infect them. Transient flora are the reason that maintaining good hand hygiene is so important. Coronaviruses also belong to this group.

In the third group are the opportunists. They only cause illness when all the conditions are just right, but they also often do little damage. They occur in places they normally don't belong, such as when large amounts of candida suddenly appear in the gut as a consequence of an unhealthy diet or an unnecessary course of antibiotics.

To keep all these different kinds of flora healthy, it's important that we do as much as we can proactively to keep our immune system in optimum condition. In this book, I explain the hows and whys of all that. Consider this chapter as a dive into the terminology. You don't need to learn everything by heart. We'll start by taking a little look at the creatures we live with all the time, then explore how our immune system deals with them and what we can do to optimise those interactions. So, let's get going!

'Life is not a bowl of cherries'

Bouba Kalala

MICROBIOLOGY

1

If we are to talk in depth about immunology, we must first have a good understanding of how the microbiological world works. Microbiology is the science of life on a tiny scale. Although our immune system is always working at full throttle, it springs into action in a very ingenious and special way when our body comes into contact with the outside world, i.e. with the tiny creatures that are the object of microbiological study.

Pathology

Let me introduce you, dear reader, to the subdivisions of disease. And yes, I'm trying to impress — doctors are like that! Bandying a few exotic-sounding words about, a bit of showing off here and there with complicated medical terms. But I promise you that you will understand them all, and even be able to use them meaningfully yourself.

Pathology is the study of the causes and effects of disease or injury. In pathology, we can subdivide diseases in different ways. One way is to classify them according to the timeframe of the illness: they can be acute, chronic, or latent.

An acute illness is like a Ferrari racing by. The illness arrives quickly and then it's gone in just as short a time. The common cold is an example of an acute illness that comes and goes with no serious consequences. Sometimes, an acute illness can indeed cause a lot of damage in a short time, with possibly more-serious consequences. Sometimes, an acute illness can kill you.

Another type of illness is a chronic condition. It comes on more slowly and lasts much longer. Some acute illnesses evolve into chronic conditions, often lasting several months. Everyone probably knows someone with chronic sinusitis, in which the nose and paranasal sinuses are in a state of chronic infection. The mucous membranes that line the sinuses are continuously swollen and produce abnormally large amounts of mucus. Ironically, sinusitis sufferers often complain of having a 'heavy head', although the very purpose of the sinuses is to make the head lighter: nature makes sure our head isn't too weighty by incorporating air-filled cavities into it.

Latent illnesses are conditions that appear, disappear for a while, and then reappear. Shingles is such a disease. It's an outbreak of the herpes virus, which can live with us throughout our lives, nestled in the roots of our nerves from the first time we come into contact with it. It can also live in the nerve roots of the lips, eyes, or genitals and typically manifests itself as a skin rash and painful, fluid-filled blisters in the affected area. If the virus gets into the nerves of the lips, the result is cold sores, which can reappear whenever your defences are a little down.

Another way of subdividing diseases is by determining where exactly in the body the infection is located.

A local infection is clearly demarcated: a small cutaneous abscess, for instance, has no consequences for the rest of your body. Another name for that is a pimple, which most of us are familiar with. We all struggled with hundreds, or even thousands, of 'small cutaneous abscesses' while going through puberty.

This stands in contrast to a systemic infection, in which the disease is everywhere in your system. It often spreads through the body via the bloodstream, in which case it's known as blood poisoning, or sepsis, and can lead to a coma. Flesh-eating bacteria are a good example of this: the problem begins very locally, perhaps with a small skin lesion, which lets the bacteria into the body. When the toxins from the bacteria enter the bloodstream and begin to circulate round the body, the blood vessels widen and the patient's blood pressure falls sharply. This happens very quickly, and so a systemic infection such as this can be deadly.

Because medicine likes to make everything unnecessarily complicated, there's yet another way to subdivide diseases — namely, primary and secondary infections. A primary infection is easily explained: first you're healthy, then you catch a cold — and you have a primary infection.

You get what we call a secondary infection if you develop a second infection from the first one — in this case, it could be pneumonia on top of your cold. The pneumonia is a result of the original cold. The defences in your lungs are weakened by the first infection, making you more vulnerable to the second one. Secondary infections are a common problem among older patients.

In general, diseases progress through five stages.

The first stage is the incubation period, which is the time between the moment you come into contact with a microbe and the first appearance of symptoms. Suppose you meet someone with the disease COVID-19. You shake that person's hand, then rub your eyes, and the virus enters your body. You feel perfectly normal for three, four, maybe five days. The incubation period is the time the virus is already in your body, preparing to make you sick, without your feeling anything yet. But you are already infectious! So the virus can continue to spread, as you may shake someone else's hand before you're aware that you're ill. All infectious diseases have an incubation period.

The second stage is the prodromal phase. *Prodromos* is Greek for 'forerunner' and so refers to the early symptoms of a disease. The first indications may be minor — an initial sign that your immune system is starting to stir. You might get a mild headache, general malaise, a little muscle pain. This phase often isn't very serious, but the prodromal phase should be taken as an indication that you need to take it a little easier for a while.

Next comes the invasive phase, when the typical symptoms of a disease appear. In the case of COVID-19, that may include loss of the senses of taste and smell. General symptoms during the invasive phase of any illness are a high temperature, nausea, redness, and a cough. This is the time when the illness reaches its peak intensity. Your body is in action: on this battlefield, your immune system is fighting for all it's worth on the front line against foreign invaders.

This is followed by the decline phase. Symptoms diminish: your immune system is winning the fight against the pathogens. In this phase, you must be doubly careful to ward off secondary infections. While your immune system is fully occupied with one pathogen, another may be more able to sneak in unnoticed.

The final stage is the convalescent phase. Your body repairs damaged tissue, complete recovery is achieved, and your body returns to the condition it was in before the infection. It regains strength and you no longer have any symptoms, although you may still be infectious.

Disease Progression

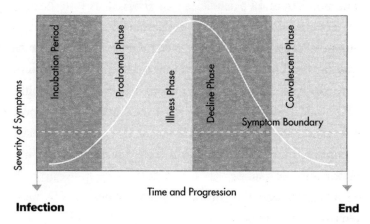

Severity of Symptoms

Incubation Period

Prodromal Phase

Illness Phase

Decline Phase

Convalescent Phase

Symptom Boundary

Time and Progression

Infection

End

But how do we actually get sick? Sometimes, it's due to the bacteria themselves: they make us sick directly with toxins. Yet our own immune system can also go into overdrive when it has to fight a bacterium or virus. Symptoms that can be caused by the immune system's springing into action include fatigue, fever, muscle pain, and a general feeling of malaise. Sometimes, an immune reaction can make us feel more ill than the pathogen itself does.

Let me zoom in to the case where bacteria make us sick directly, and the toxins involved. There are two main kinds of toxin related to bacteria. Exotoxins are poisonous substances that are produced by the bacteria and released into our bodies. Endotoxins are bacteria that are poisonous in themselves. Many bacteria have certain substances in their cell walls that are toxic for humans. The best known of these are the lipopolysaccharides, or LPS. These are fatty sugar molecules that nestle on the outside of microbes. These toxins are often present in cases of blood poisoning. The number of LPS particles can also tell us how healthy someone is. We carry a wide range of different bacteria in our gut, known collectively as our microbiome (more on that later). The less healthy the make-up of that microbiome is, the more LPS there will be in your blood. Higher baseline LPS values are associated with a greater risk of chronic infection and a general lack of wellbeing, because the body is constantly in a state of mini-blood-poisoning. Hence the importance of pampering the little beasties in your belly with healthy food. Forgive me, but that's a message I'll repeat 389 times in this book.

The way your immune system reacts to an invader can also make you feel ill. This happens with coronavirus: it creeps into the alveoli of your lungs, where your immune system becomes activated. All sorts of cells and substances get involved, and some of the cells don't survive the battle. We'll get into the details of that later, but what's important here is the pus that remains after immune cells die. The alveoli become filled with pus and can no longer work the way they should. So your body gets sick — not because of the virus per se, but because of your body's defensive reaction to it: because of the pus filling your lungs. This is the infamous ARDS, or acute respiratory distress syndrome. Someone whose body is healthy and whose lungs work well will probably not be bothered too much by this. But for someone with pre-existing conditions, such as obesity, diabetes, or other lifestyle diseases, and reduced lung function, this situation is potentially very dangerous.

Microbiology Is Everywhere

You may ask why I'm going into so much scientific detail. Microbiology has a great impact on human health. Unfortunately, that fact has been made more than clear to us recently. We had all conveniently ignored the fact that infectious diseases can have serious consequences for us, even in the 21st century. But nature always wins out. We can't simply ignore the tiniest organisms on our planet. Indeed, we need them, since they are at the beginning of the food chain, they are in our gut, we use them to make food and drinks — just think of the microbiological processes involved in beer brewing, bread baking, or cheese making. And in agriculture, it's thanks to microbiology that we can produce disease-free crops on a large scale.

But who are the tiny inhabitants of the microbiological world?

Microorganisms are invisible to the naked eye. We distinguish between two types: eukaryotes and prokaryotes. *Karyon* is the Greek word for 'kernel', and *eu* is Greek for 'good' and, by extension in this case, 'present'. So a eu-karyote cell is one that has a kernel or, in microbiological parlance, a nucleus. Pro-karyote cells do not have a nucleus.

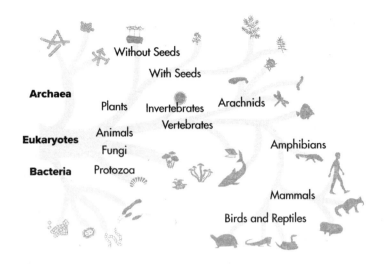

As prokaryotes, bacteria — and another unicellular form of life, archaea — have no cell nucleus and their DNA floats around inside the cell. At some point, evolution 'decided' it could be useful to keep DNA together in a nucleus, and that was the beginning of the eukaryotes. In a subsequent evolutionary stage, cells that contained a nucleus started to stick together, and that was the beginning of the fungi and the plants. Further refinement led to the development of more complex life forms, and the idea of combining cells with different functions and capabilities in a living organ, and later an organism. This explains how animals evolved. Yet all living organisms can still be divided into two groups — those whose cells have a nucleus and those whose cells don't.

Viruses don't fit into this classification scheme. In fact, viruses are proteins that float around freely. Even scientists aren't yet sure whether they can be considered life forms or not! They come in different shapes and sizes, as the following illustration shows.

Virus Structures

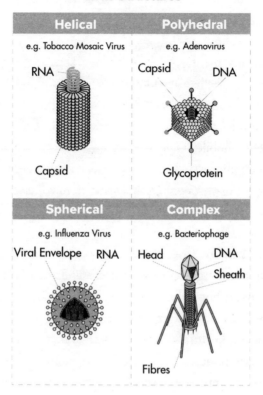

Helical	Polyhedral
e.g. Tobacco Mosaic Virus	e.g. Adenovirus

RNA
Capsid

Capsid
DNA
Glycoprotein

Spherical	Complex
e.g. Influenza Virus	e.g. Bacteriophage

Viral Envelope   RNA

Head   DNA
Sheath
Fibres

Bacteria Structures

Cocci	Bacilli	Others

Diplococci   Streptococci

Streptobacilli

Vibrios

Tetrad

Flagella

Spirilla

Staphylococci   Sarcina

Spore-forming

Spirochaetes

The wonderful world of microorganisms can be categorised into six groups:

- As already mentioned, bacteria are super-tiny creatures: nothing more than a cell without a nucleus. This means that they are usually a thousand times smaller than cells that do have a nucleus. There are different types of bacteria, and some are mobile, while others aren't. They get their energy through photosynthesis — the conversion of sunlight into energy — or through chemosynthesis, in which the energy comes from the conversion of chemical substances. They can also use both systems. What ingenuity!
- Algae can be unicellular or multicellular, and are eukaryotes. Algae can also be mobile or immobile, and they use photosynthesis, but they do not frequently cause disease. Blue-green algae can cause skin irritation and gastrointestinal problems if found in swimming pools and ponds.
- Protozoa were the first, single-celled animals. Malaria, caused by various *Plasmodium* species, is a protozoan infection (mosquitos are only the carriers of this devastating disease).
- Fungi, or moulds, are eukaryotes, so their cells have a nucleus. They can be multicellular; in their single-celled form, we call them yeasts. Moulds are not mobile, and they gain energy both from photosynthesis and from the chemical substances in the organisms they grow on. For those who want to improve their language skills: the Greek for 'mould' is *mykos*, which is why the medical word for a fungal skin infection is mycosis. That tends

to happen when there's an imbalance in your normal flora, caused by a course of antibiotics or an unhealthy lifestyle. You're surely familiar with the mould that grows on bread, so you also know that sugar provides an ideal breeding ground for these creatures.

- In the fifth group are the viruses, which aren't really capable of anything without using other cells. They can survive in the air for a short while, in which case we call them airborne, but they require a host cell to survive for any longer. They consist of nothing but genetic material surrounded by a protein shell. They hijack the nucleus of a host cell by inserting their own genetic material into it and thus reproducing. Every virus has a preferred type of host cell. The common-cold virus, for example, sneaks into the cells of the respiratory tract, while the norovirus nestles in the cells of the gut, and the herpes virus prefers our sensory nerve cells. Viruses can't move under their own steam, but they can, for instance, ride on your spittle to reach another host, where they infiltrate more cell nuclei and begin to divide. The original cells become sick and die, and anywhere from ten to a thousand more viruses can then spread, and also escape into the environment.

- The sixth and last group contains the helminths. This group includes parasitic worms such as flatworms, roundworms, etc. As pathogens, these can be actual little worms, but also their larvae or eggs.

Microbiology Has Rock Stars of Its Own

One of the founding fathers of microbiology as a science was Antonie van Leeuwenhoek. He was a draper and biologist, but he was also the inventor of the first microscope. The name comes from the Greek *mikros*, 'small', and *skopein*, 'to look at' — so, it's an instrument for looking at small things. This technology allows us to see things that are invisible to the naked eye. The optical microscope, using lenses to magnify objects, was an important development in the science of microbiology. Now, scientists also use electron microscopes to study even smaller organisms.

The physicist Robert Hooke also deserves a special mention here. Around 1665, while studying the detailed structure of cork, he discovered that all living things are made of cells.

For a long time, people believed living organisms arose out of non-living matter. It was believed that maggots appearing on a piece of meat did so on their own. The theory was known as 'spontaneous generation'. Francesco Redi was the first scientist to disprove the theory and to postulate that no maggots can appear on meat unless flies have previously deposited their eggs on it. Of course — as was to be expected — people at the time refused to believe him, and the entire scientific world baulked at his theory. But the truth will always out!

John Needham and Lazzaro Spallanzani are two more scientists who investigated this theory. They showed that spontaneous generation did occur, but only in the presence of air. They experimented by placing meat in sealed and unsealed jars. Meat that was exposed to the air changed, with all manner of things beginning to grow on it.

Building on this work, the famous scientist Louis Pasteur discovered that many illnesses could only develop in the presence of microorganisms in the air or in dust. He was able to support this 'germ theory' with his discovery of the process of pasteurisation, in which he briefly cooked meat and observed that no microbes grew on it afterwards. He extended use of the process to other products, such as milk, which could be stored for longer once they were germ-free. Pasteur was able to show that plants and foodstuffs do not spoil due to spontaneous generation. (Incidentally, Pasteur also created the first rabies vaccine.)

One extremely important member of our microbiological all-star panel is Robert Koch. Around the year 1870, he proposed the very interesting theory that certain microbes cause certain diseases — the germ theory of disease. He worked mainly on anthrax, an infectious disease that causes blisters and can lead to pneumonia or destroy other body systems. Koch's experiments enabled him to identify the pathogen *Bacillus anthracis*, by injecting it into test animals to see if they contracted the disease. He also discovered the bacterium that causes tuberculosis, which was therefore originally known as Koch's bacillus. This work earned him the Nobel Prize in Medicine. He published what became known as his four postulates, which forever changed our understanding of microbiology and the capacity of microorganisms to cause disease:

- the pathogen is present in every individual suffering from the disease
- the pathogen must be isolated and grown in pure culture
- a healthy animal gets sick if injected with the pathogen
- samples of the pathogen can be extracted from the deliberately infected animal.

Then, in the 19th century, one Hans Christian Gram discovered by chance that there are two different kinds of bacteria. He sprayed some dye on the glass plates he used for looking at bacteria, and saw that some of the microbes turned red, while others turned blue. This was the origin of Gram staining, a technique still used today to classify bacteria into categories. The course of a disease and the choice of treatment for an infection differ depending on whether you are dealing with blue, Gram-negative bacteria or red, Gram-positive bacteria. Thus, with his Gram's fairytale, this Hans Christian had a positive influence on today's modern medicine.

Bacteria in Detail

Enough about humans, now it's time for the bugs to enter the limelight!

Prokaryotes always consist of three parts:

- some sort of envelope, which is the cell wall or membrane
- something inside
- one or more appendages.

The structure of a bacterium determines not only what shape it is, but also how it can move around, how fast it can go, and so on. The structure of the cell wall is important because it's partly responsible for determining the mechanisms of the disease it causes, how the immune system reacts to the bacterium, and which medication will successfully kill it — and which won't. There's a kind of slimy layer — or, to use the fancy medical word, glycocalyx — on the wall of a cell. It's made up of sugar and protein molecules. This slimy capsule protects the bacteria from dehydration and external attacks, and determines the extent to which a bacterium can make us sick. Just consider why dental plaque sticks to your teeth. The sugars and proteins in the glycocalyx of the bacteria in the plaque love to cling to the enamel of our teeth.

The DNA of the bacterium is stored in the ingenious envelope formed by the cell wall and the cell membrane. The cell wall sustains the bacterium's survival. It also contains a few ribosomes, which can be described as the untanglers of our DNA. Ribosomes are like 3D printers that synthesise genetic information into proteins. Some bacteria also contain the lipopolysaccharides (LPS) mentioned earlier, which are released in the gut and can be absorbed into the blood. Too many lipopolysaccharides can cause illness and chronic infection.

Eukaryotes have a more sophisticated cell structure, with a cell nucleus and other appendages. Some of those

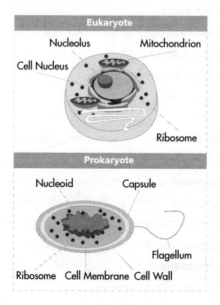

appendages help the cell move, but they can also help it absorb nutrients.

Having a nucleus, eukaryotes are more complex in structure, with many more little elements in the cell fluid and a nucleus containing genetic material and ribosomes, all surrounded by the endoplasmic reticulum — if you'll pardon my language. This is a network of membranes that surrounds the nucleus and is connected to it. This is how the genetic material from the nucleus is transported to the Golgi apparatus, where the freshly made proteins gather to be released into the cell. Eukaryotic cells also contain mitochondria, which are their powerhouse. (And this is pretty interesting, because mitochondria evolved from primitive bacteria that chose to keep eukaryotic cells alive, in a symbiotic relationship.)

The Name Game
All organisms on earth fit into a system of classification. This means we can identify any organism, determine which family it belongs to, and analyse its properties.

Take human beings as an example:

- kingdom: animals
- phylum: vertebrates
- class: mammals
- order: primates
- family: hominids
- genus: *Homo*
- species: *H. sapiens.*

The categories for microorganisms are very similar. Here's the microbe that causes syphilis:

- kingdom: bacteria
- subkingdom: Gram-negative bacteria
- phylum: spirochaetes
- class: spirochaetes
- order: Spirochaetales
- family: Spirochaetaceae
- genus: *Treponema*
- species: *T. pallidum.*

We usually refer to organisms by their genus and the specific name identifying their species — a little bit like us using first and last names. So: *Servaas Bingé, Staphylococcus aureus, Treponema pallidum.* If you know a bacterium's first and last names, you can deduce the disease it causes: *Treponema pallidum* causes syphilis; *Staphylococcus aureus* causes skin infections. (According to the latest scientific findings, *Servaas Bingé* isn't known to cause any infections.) Some other examples are *Plasmodium malariae, Neisseria gonorrhoeae* (colloquially known as the clap), *Bordetella pertussis* (whooping cough), *Streptococcus pneumoniae* (pneumonia), and *Chlamydia trachomatis.*

Pathogen (Virus, Bacterium, Fungus, Parasite)	**Disease**
Aspergillus genus	Aspergillosis
Bacillus anthracis	Anthrax
Mumps orthorubulavirus	Mumps
Bordetella pertussis	Pertussis (whooping cough)

Borrelia burgdorferi	Lyme disease
Candida genus	Thrush
Chlamydia pneumoniae	Pneumonia
Chlamydia psittaci	Psittacosis (parrot fever)
Chlamydia trachomatis	Chlamydia
Clostridioides difficile	Diarrhoea and colitis
Clostridium botulinum	Botulism
Clostridium tetani	Tetanus (lockjaw)
Coltivirus genus	Colorado tick fever
Corynebacterium diphtheria	Diphtheria
Coxiella burnetii	Q fever
DENV-1, DENV-2, DENV-3, DENV-4 (*Flavivirus* genus)	Dengue fever
Ebolavirus genus	Ebola
Enteroviruses (incl. Coxsackie A virus)	Hand, foot, and mouth disease
Epstein-Barr virus	Mononucleosis (mono, glandular fever, the 'kissing disease')
Escherichia coli	Urinary tract infections, intestinal infections, other infections
Giardia duodenalis	Giardiasis
Group A streptococcus	Scarlet fever
Haemophilus influenzae	Hib
Helicobacter pylori	Gastritis and stomach ulcers
Hepatovirus A	Hepatitis A
Hepatovirus B	Hepatitis B
Hepatovirus C	Hepatitis C
Herpes simplex virus 1 and 2 (HSV-1 and HSV-2)	Cold sores, shingles, genital herpes
HIV (human immunodeficiency virus)	AIDS (acquired immune deficiency syndrome)
Human papillomavirus	HPV, cervical cancer
Human herpesvirus 6 and 7 (HHV-6 and HHV-7)	Exanthema subitum (roseola, sixth disease)
Human parainfluenza viruses (HPIV)	Upper respiratory infections
Legionella pneumophila	Legionnaire's disease
Listeria monocytogenes	Listeriosis

Measles morbillivirus	Measles
MERS-CoV	Middle East respiratory syndrome (MERS)
Molluscum contagiosum virus (MCV)	Molluscum (water warts)
Mycobacterium leprae	Leprosy
Mycoplasma pneumoniae	Pneumonia
Neisseria gonorrhoeae	Gonorrhoea (the clap)
Neisseria meningitidis	Meningitis
Norovirus genus	Gastroenteritis in children and newborns
Orthomyxoviridae family of viruses	Flu
Parvovirus B19	Erythema infectiosum (slapped cheek, fifth disease)
Pediculus humanus capitis	Head lice, nits
Plasmodium genus	Malaria
Poliovirus	Poliomyelitis (polio)
Pthirus pubis	Pubic lice (crabs)
Rabies lyssavirus	Rabies
Rhinoviruses and coronaviruses	Colds
Rickettsia genus	Typhus
Rickettsia prowazekii	Epidemic typhus
Rotavirus genus	Gastroenteritis
Rubivirus rubellae	Rubella (German measles)
Salmonella genus	Gastroenteritis (food poisoning)
Sarcoptes scabiei	Scabies
SARS-CoV-2	COVID-19
Shigella genus	Shigellosis
Staphylococcus genus	Staph
Streptococcus genus and *Staphylococcus aureus*	Cellulitis
Streptococcus pneumoniae	Pneumonia
Streptococcus pyogenes	Skin infection and sepsis
Toxoplasma gondii	Toxoplasmosis
Treponema pallidum	Syphilis
Trichomonas vaginalis	Trichomoniasis (trich)
Trichophyton genus	Tinea (athlete's foot, ringworm)
Varicella-zoster virus (VZV)	Chickenpox and shingles

Vibrio cholerae	Cholera
Yellow-fever virus (*Flavivirus* genus)	Yellow fever
Yersinia pestis	Plague
Zika virus (*Flavivirus* genus)	Zika fever

How Can We Keep Bacteria Under Control?

There are various ways we can keep bacteria at bay using methods of physical control, including sterilisation. True sterilisation is achieved by heating to at least 121 degrees Celsius (250 °F) for at least an hour, which is enough to kill any organism. But there are also many bacteria that are killed by more ordinary temperatures — just think of the process of pasteurisation. Some bacteria, such as the *Legionella* that live in our water pipes, can only survive temperatures of up to 70 °C (160 °F). If the water in the pipes is heated to a sufficiently high temperature, those pathogens can be stopped from spreading.

On the other hand, cold stops most microbes from growing, as it robs them of their responsiveness. Only *Listeria* bacteria are able to survive in your household freezer. These bacteria are found mainly in products that are kept frozen for a long time and then eaten without being heated. Risky foodstuffs include cheese, fish, and animal products.

Electromagnetic or ionising radiation can also be used to kill bacteria. These sterilisation methods are increasingly being used in hospitals and operating theatres.

Bacteria can also be contained using chemical control methods, involving soap, alcohol, oxygenated water, and so on. While experimenting on bacteria, a certain Mr Lister (yes indeed, the man who gave his name to those *Listeria* bacteria in your cheese) used phenol to clean his laboratory apparatus and discovered that the chemical also killed bacteria. Other antimicrobial products or surfactants (like soap) don't kill bacteria but rather wash them away.

Currently, we're all expected to wash and disinfect our hands 13 times a day, so I'm sure you're asking yourself which is better: washing your hands or using hand sanitiser?

The answer is that you should first wash your hands with soap, which removes any microscopic dirt particles. Bacteria and viruses stick to us because of the oily substance on our skin called sebum. Soap is a degreasing agent and helps wash away the sebum, together with the bacteria and viruses stuck to it. You can make sure you get rid of everything by using the special handwashing technique that we have now all become so familiar with:

Instructions for Hand Washing

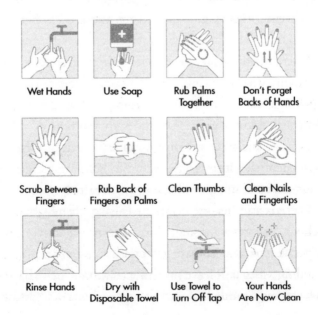

Wet Hands	Use Soap	Rub Palms Together	Don't Forget Backs of Hands
Scrub Between Fingers	Rub Back of Fingers on Palms	Clean Thumbs	Clean Nails and Fingertips
Rinse Hands	Dry with Disposable Towel	Use Towel to Turn Off Tap	Your Hands Are Now Clean

However, your hands, or whatever body parts you are cleaning, are not yet sterile. Sterility is achieved when a surface is completely germ-free, as with medical apparatuses or instruments in an operating theatre. Any remaining pathogens can be killed by using a disinfectant or rubbing alcohol after washing with soap and water. We no longer use real phenol, as Lister did, since it causes skin damage to the hands, but we do use solutions such as chlorhexidine, which is used to disinfect hospital equipment.

In the past, oxygenated water (such as hydrogen peroxide) used to be used to rinse and clean wounds. The oxygenated water causes the tissue to fizz: a sign that all the bacteria are being eradicated from the wound, as the oxidation destroys their cell walls. Older readers might remember using Mercurochrome, a mercury-based product to treat wounds — which we no longer use, as we now know that mercury is quite toxic. But it was effective, nonetheless!

Today, the best way to treat a wound is to wash it thoroughly with soap and water, then to disinfect it as well as possible with a chlorhexidine-based product, and not to cover it. Wounds heal more quickly when they are left open to the air, because that allows the acidity of the skin to remain at the optimum level, although, on the downside, leaving a wound uncovered exposes it to reinfection.

There is another important reason for keeping bacteria in check: to preserve food. We can heat up our food or refrigerate it. A foodstuff's level of acidity can also be changed by adding vinegar. That kills any pathogens, and the food can be stored for much longer. Salting also makes food more acidic, thus killing any microbes on it. It doesn't sterilise the food, but creates a hostile environment for bacteria.

And then there are artificial preservatives, of course ... I could enter into a whole discussion on them, but that would likely lead us too far away from our current topic. Some of those preservatives are added to bread to stop it from going mouldy. Sodium nitrates ensure that meat remains free of botulism bacteria. I could mention many more examples — just look at the list of ingredients of any food product, especially those represented by numbers.

Are all numbered food additives preservatives? No, some are flavour enhancers, colourants, or emulsifiers (which make it easier to mix ingredients). Are all artificial preservatives automatically bad for you? No, many are completely harmless. The substance designated E300 under European food-labelling laws, for example, is ascorbic acid, or ordinary vitamin C. But there are also some that are really to be avoided. Explanatory lists of food additives can be found on the internet.

If in doubt, remember my motto: the more natural your diet is, the healthier you will be. Eat as many things as possible that grow naturally!

One of the greatest dangers of preservatives is that they can trigger problems such as skin irritations, breathing difficulties, and bronchitis. If you have a medical problem that defies explanation, try cutting artificial additives out of your diet as much as you can. It may not be detectable in a blood sample, and such reactions are of course dependent on the dose, but if you've tried everything else, this trick can sometimes help.

Preservatives can sometimes cause real problems in children. Again, there is unfortunately no one-to-one measurement that can be made, but there is increasing evidence for a link between artificial or industrially processed food and behavioural problems in children. The principle behind this hypothesis is that such artificial substances influence low-level inflammation. The growing nervous tissue of our little ones also becomes inflamed — a condition known medically as neuroinflammation. I suspect that the concept of neuroinflammation will have a place in our understanding of many diseases of the nervous system in the coming years.

Consuming too many artificial preservatives over a long period of time is unhealthy and can be dangerous. It can even exacerbate certain conditions, such as heart disease in older patients, or obesity. Unfortunately, this is still a very young science and precious little research has been published on the subject.

The final method of keeping the bugs at bay is through chemotherapy. We tend to associate *chemotherapy* with cancer treatments, but, in fact, the term refers to the entire range of medications that are toxic to pathogenic cells, but (usually) not to patients themselves.

Antibiotics

Antibiotics are one group of such medicines. Their name gives a clue to their function: 'against life' (*anti-bio*). There are different types of antibiotics, and they work in different ways. Some destroy the bacteria's cell wall; others disrupt the deeper cell membrane, causing the bacteria to burst.

Doctors prescribe broad-spectrum antibiotics when they are unable to immediately identify the type of bacteria they're dealing with, and they want to be sure they don't miss any pathogens. We now try to use broad-spectrum antibiotics as sparingly as possible, because when bacteria are exposed to them often enough, they develop resistance to the drugs, which then cease to be effective. This resistance is a major problem in hospitals: some specific bacteria are quite resistant to the classic medications. Fortunately, people have become far more aware in recent years of the dangers of drug resistance and of the dangers of overusing antibiotics or prescribing them indiscriminately. The use of broad-spectrum antibiotics can't be completely avoided, but we really must restrict their application to cases where it's certain that the cause of the infection is bacterial. Antibiotics have no effect against viruses and will not shorten the duration of a viral disease.

Another problem with antibiotics is their side effects. They can be toxic for some people — penicillin, for instance, can trigger an allergic reaction in some cases.

A third barrier to the use of antibiotics, especially broad-spectrum antibiotics, is the fact that they disrupt the body's normal microflora. Firing a broad shot at everything that moves destroys the body's own ingenious system of microorganisms, which perform some very important functions. Destroying the microbiome in the gut can give free rein to fungal infections, which can develop into chronic conditions. When treatment with antibiotics is required, it's wise to take some precautions. These can include the use of probiotics, or trying to include as many prebiotic foods in your diet as possible while you're on the course of antibiotics, to protect yourself against undesirable side effects. Prebiotic food creates the right climate in the gut to grow beneficial bacteria, as do foods that are high in fibre, as well as kombucha, kefir, and even humble yoghurt.

Traditional food is part of traditional health, and I can say that I'm now noticing a change in attitude among the general public ... we are better informed, pay more attention to the consequences of our lifestyles, and take prevention seriously. Society is moving in the right direction: people are more critical and no longer take whatever the doctor prescribes without question while ignoring their own diet.

But let's get back to broad-spectrum antibiotics. Why do they still exist, if they are so bad for us? Well, waiting to treat a disease until the cell cultures that tell us which bacteria are responsible for it to grow can mean it's already too late. Luckily, better diagnostic techniques for rapid pathogen identification are emerging. This will become more and more significant in future, as we become able to treat diseases in an increasingly targeted way. But a medical condition can potentially be caused by a number of different bacteria. There are three or four different germs that cause throat infections, for example. If, during the early stages of the infection, when it's still unclear which pathogen we're dealing with, we fire our weapons at one of the four possible causes, it may be the case that we're leaving three others untreated, and that doesn't make for optimum treatment of the sore throat.

The moral of the story is: antibiotics are often un-necessary, but sometimes they are vital. Governments and medical institutions have invested a lot of time and energy in making doctors aware of their responsibilities. We have been taught the beneficial effects of antibiotics, as well as their side effects, and now it's up to us to use them responsibly.

How Viruses Work

On to the next lesson in this microbiological crash course — on viruses!

What's in a Name?

The word *virus* is Latin and means 'poison'. That in itself is bad news enough! I already mentioned, albeit in less fancy terms, that viruses are obligate intracellular parasites. In real terms, that means a virus needs a host and can't reproduce without coming into contact with another cell. Are they alive? On the one hand, they have a metabolism and they reproduce ... so that means they're alive, right? On the other hand, they can't survive without a host cell ... so ... they're not alive? We simply don't know, but it doesn't really matter. What we do know — all too well, unfortunately — is that viruses can have devastating consequences.

Viruses come in all shapes and sizes, and can be categorised in several different ways:

- According to size. Some are big; others are small. In general, viruses are around a quarter to a tenth the size of bacteria.
- According to their nucleus. Viruses consist of a protein coat around a nucleus of DNA or RNA. The protein coat is called the envelope, as can be seen in the illustration. The genetic material, or the virus's genome — its DNA

or RNA — is contained within the capsid. I assume you're familiar with DNA (deoxyribonucleic acid), which contains the genetic code that records hereditary characteristics, such as eye colour, skin type, and so on. For those hereditary characteristics to be expressed, the body needs to convert that DNA into proteins. All bodily functions are in fact protein reactions — just think of muscle protein and enzyme reactions. However, DNA can't be converted directly into proteins, which is why RNA (ribonucleic acid) is necessary. It activates the cell's protein factories, the ribosomes. Viruses either contain DNA, in which case they have to invade the nucleus of the host cell and interfere with its DNA, or they contain RNA, in which case a virus doesn't have to take that detour and can act on the cell's protein factories directly.

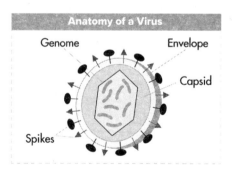

I already mentioned that viruses need a host cell. They use their protein coat, or envelope, to attach to the wall of the host cell and then thrust their genetic material into its nucleus. A virus that surreptitiously enters a host cell in this way can quickly begin making hundreds of thousands of copies of itself. The kind of cell a virus needs to replicate depends on what type of virus it is. One that requires a canine cell is harmless for humans. And, by the same token, viruses that require human host cells may be completely harmless for dogs.

The Bat and the Pangolin

I say 'may be' because, unfortunately, viruses can mutate. This is what happened to the novel coronavirus. It probably originally required bats to reproduce. One day, the virus mutated, making it feel just as at home in humans. We may never be able to reconstruct this sequence of events completely, but it's supposed to have something to do with bat urine and a pangolin. I must admit that I wasn't present at the time, so I can't say anything for certain. But it's highly likely that the source of the virus was an animal, after which it mutated, making it able to be passed from human to human. On a purely evolutionary level, this happened by chance and could have turned out completely differently.

A Day in the Life of a Virus

So, what actually happens when you get infected with a virus? Let's start by following in the footsteps of the coronavirus. When someone who's infected with the virus coughs, a myriad little corona particles are propelled into the air. Those coronaviruses, equipped with a machinery adapted to cells in the respiratory tract, land on your skin and your mucous membranes. They can't affect skin cells, but they are superbly equipped to attack mucous-membrane cells. After penetrating such a cell, the virus takes off its protein coat. It then integrates its genetic material into the mucosal cell and starts to replicate itself. Incidentally, this is also how gene therapy works: innocent viruses are used as carriers to insert genetic material into the patients' cell nuclei.

But back to our corona infection. The host cell realises that all is not well and starts to defend itself, but the virus works quickly and efficiently. In next to no time, it produces so many virus particles that the original cell breaks down, and the hundreds of components form new viruses. They are then released into your body — and each one goes on to do the same to a neighbouring cell. The speed at which that takes place differs from virus type to virus type.

By this time, your immune system will have sounded the alarm. In the case of a common-cold rhinovirus, your immune system goes on the attack, and any cells containing the virus are shrivelled up and rendered harmless. The herpes virus, however, likes to put its little feet up and relax, hidden safely inside its host cell. It can't move from there as long as your body remains alert, but, as soon as that vigilance weakens, due to fatigue or another infection, the immune system forget about the virus, giving it the freedom to reproduce to its heart's content. The medical term for this process is reactivation.

Examples of DNA viruses:

- Herpes viruses, of which there are several subfamilies. This group includes the viruses that cause cold sores, genital herpes, and chickenpox. When a young child is infected with the varicella virus, it develops into chickenpox all over the body. This is because children's defences are not yet well developed. In adults, the virus nests in a specific part of their nervous system, and immunity is only compromised in one part of the body, causing cold sores on the lips, herpes blisters in the genital area, or shingles elsewhere.

- The Epstein-Barr virus. This is actually another herpes virus, and can cause throat infections and mononucleosis in young adolescents. In older adults, it can be associated with various types of cancer, such as Hodgkin's disease.
- *Variola* viruses, which cause smallpox. Thanks to the World Health Organization's exhaustive vaccination strategy, the world was declared free of smallpox in 1980.
- Parvoviruses, like the children's illness known as the fifth disease, cause high temperatures and rashes, but they pass very quickly.

Examples of RNA viruses:

- Retroviruses, which suppress the immune system, leaving us more susceptible to other conditions. HIV, which can lead to AIDS, is probably the best-known RNA virus.
- Flaviviruses cause yellow fever, dengue fever, and zika fever.
- The family of enteroviruses includes polio, which leaves children paralysed but has been largely eradicated thanks to vaccinations. Unfortunately, it still crops up from time to time. Here I must make special mention of the Bill & Melinda Gates Foundation, which continues to make extensive efforts to suppress the virus worldwide.
- Rhinoviruses cause the common cold.
- Influenza or flu viruses. The flu is a special case. Very

often, two different but similar flu viruses will merge inside a cell and create a new virus type. Using a term that sounds as if it comes from one of the *Matrix* movies, this is known as antigenic shift and means the virus can mutate at lightning speed. This is why the flu virus that causes each annual epidemic is slightly different from the flu virus that caused the previous one. It explains why we need a different vaccine every flu season.

- The rabies virus gives you rabies. It's passed on by bites from dogs infected with the disease, and is a virus that can be transmitted from a canine cell to a human one.
- Measles, mumps, and rubella belong to the group of paramyxoviruses (all these strange long words must be starting to make you feel like you've ended up at Hogwarts).
- The Ebola virus, from the family of the filoviruses, causes fatal internal bleeding and is extremely contagious. We occasionally see outbreaks of this virus in Africa.
- The well-known rotavirus may also ring a bell: it causes severe diarrhoea and stomachaches, especially in small children. An orally administered vaccine against this disease was launched on the market a decade ago.

Some of these viruses are becoming increasingly associated with cancer. Most cancers arise due to genetic mutations in the patient's DNA. Those mutations are caused by chemicals, nicotine, pesticides, radiation, you name it ... So, the genetic material within a given cell undergoes a change. That's exactly the way viruses work: by disrupting the code in the cell. Because the biochemical mechanism is the same, a viral infection can lead to uncontrolled growth — either benign or malignant — of the cell, which can eventually become tumorous. We know that the Epstein-Barr virus can cause liver cancer. The human papillomavirus can be caught during a first sexual encounter, and in girls it can cause cervical cancer. Boy don't come away unscathed, either: they can develop certain kinds of skin cancer. There are now vaccines against many types of human papillomavirus — a cure for (one kind of) cancer.

If reading this far has given you an irresistible urge to rush to your doctor for an all-round check-up, don't panic. The feeling is normal (ha ha). All medical students go through this phase, thinking that they're at death's door. It's so common, it even has a name — medical students' disease, or second-year syndrome.

Now I want to introduce the final player in this little virus story, the prion.

Do you really have to, Doctor? You just spent a whole chapter bombarding us with difficult-sounding words ...

Well, yes, I do have to. Prions are virus-like bodies that lack any genetic material, meaning they consist only of a viral protein coat. Not very important then, you might think, but all those who read to the end of this sentence will exclaim a loud 'aha!' — mad cow disease, also called Creutzfeldt-Jakob disease, is one condition caused by just such a prion.

Fungi and Friends

Before you run to the doctor's with an imaginary case of one of the diseases that you contracted while reading the section above, let's briefly discuss some other tiny creatures, which are neither bacteria nor viruses: the fungi. That's nothing more than the Latin word for mushroom, but it refers also to moulds and yeasts. One of the main characteristics of fungi is that they can't move under their own steam. But they do have cells and cell nuclei, and they can spread within a host organism.

Of course, we all know and love mushrooms — 'normal' fungi that grow in nature and can be poisonous, while others can be absolutely delicious to eat.

Fungi form an important link in the chain of decay and renewal of material. They play a vital part in composting and recycling, but they sometimes have parasitic qualities, too. Just like bacteria, fungi can produce toxins, which in this case are known as mycotoxins. A disease caused by a fungus is known as a mycosis, by the way — sorry to disappoint any readers who thought the part with the difficult words was over. I already mentioned this earlier, but repetition is the mother of all knowledge, right? ;-)

Most fungi are multicellular, spread by shooting out growth filaments, and reproduce on their own, either by simply growing or by spore dispersal. Think of what happens when you blow on a mouldy orange. The powder that comes off it is composed of the spores released by the fungus. If those spores settle on a suitable surface, the fungus will begin reproducing.

Yeasts are unicellular fungi and are used in the production of bread, wine, and beer, due to their ability to ferment carbohydrates and turn them into alcohol and carbon dioxide. Some yeasts occur naturally in your gut flora but can cause illness if they start to accumulate in large numbers — candida is one such yeast.

Pathogenic moulds whose names might ring a bell include penicillin — the fungus from which Mr Fleming was able to extract a drug that can kill bacteria — and *Trichophyton* — the cause of athlete's foot, a fungal infection of the skin between the toes.

Another group of microbiological organisms that I'd like to share with you is the protozoa. Why do you need to know about them? If the worst comes to the worst, they can plague your immune system no end — these are true parasites. When I mentioned malaria earlier, what first sprang to your mind? Probably mosquitos, right? But mosquitos are not the bad guys here, they're just the delivery boys (well, girls, actually) for the pathogenic protozoan *Plasmodium vivax*. Malaria is still a very deadly disease. The name comes from a combination of two Italian words, *mal* and *aria*, together meaning 'bad air'. And it's accurate, as the breath of people with malaria has a very specific smell to it ...

Another well-known type of protozoan is *Toxoplasma gondii*, which can cause the disease toxoplasmosis. Pet cats can carry *T. gondii* without getting sick themselves; the two creatures live together in a mutualistic relationship. But, in humans with toxoplasmosis, the relationship is parasitic, and dangerous for pregnant women.

The first protozoan ever discovered was *Entamoeba histolytica*, which was found to cause severe diarrhoea in people after they drank contaminated water. The unfortunate victims described a real 'fire in their belly'.

The very last group of baddies are the worms, or helminths. Admit it, with a name like that, they sound like they could have escaped from Tolkien's *Lord of the Rings*.

Strictly speaking, helminths are animals. They live in total freedom: they can live as parasites, but they don't need to do so to survive. Types of worms you probably already know include flatworms, roundworms (nematodes), and tapeworms. The latter live in our guts, while flatworms can enter the bloodstream. Luckily, they've become very rare in our parts of the world, thanks to better hygiene, but in less developed regions they remain pretty common, unfortunately.

They used to say in the Middle Ages that tapeworms could be extracted through the mouth by first depriving the patient (and thus the worm) of food for an extended period, then luring them out with coffee, milk, or strong beef tea ... best administered by an 'old wife', I think, because the whole thing is nothing but an old wives' tale! Nowadays, there are medicines to get rid of tapeworms.

Roundworms inhabit salt water, fresh water, or the soil, and can parasitise both plants and animals. The most famous of them is called *Wuchereria bancrofti*, which settles in the lymph nodes and blocks the lymph channels, causing the affected body part to swell — hence the name *elephantiasis*. Just google it to see ... (Or maybe don't ...)

Well, that's about enough microbiology for now, don't you think?

'A belly laugh increases the ability
of your immune system
to fight infections'

Elizabeth Taylor

IMMUNOLOGY

2

I've now described in relatively broad strokes which microbes can make us sick, how they do that, and how we can try to control them. Am I trying to make an immunologist of you? No, certainly not. What's more, I myself don't pretend to be an expert in the field. But by giving you an insight into how immunity works, I hope to help you take in and make sense of all the information we're being bombarded with right now. My aim is to equip you with the necessary tools and knowledge to make up your own mind. In this chapter on immunology, I'll tell you how your immune system is organised, what happens inside your body when you get sick, and which players are sent into battle to defend against pathogenic germs.

What does the immune system look like? Unlike other systems of the body, I can't just show you a diagram to illustrate it. After all, your immune system is spread out through your entire body, and uses various different components to defend you. Let's start with a simple breakdown of the body's three lines of defence.

The Front Line

The body's front line of defence consists of structural and physical barriers. You can think of it like a medieval castle surrounded by several physical barriers to ward off invaders, such as the hill the castle stands on, the moat around the castle, and, beyond that, a ten-kilometre-wide forest surrounding the castle. Your body is protected by a similarly inhospitable area 'surrounding the castle': your first system of defence is your skin and the mucous membranes in your eyes, your nose, and your digestive tract, which runs from your mouth to your anus.

Skin and mucous membranes are exposed to the outside world and are structured with closely packed cells so that invaders — i.e. pathogens — have a difficult time getting through. For extra protection, there's a layer of dead cells on the skin and mucous on the mucous membranes. The body also produces a substance called keratin. The cells of the skin's keratin layer contain very little water, making them virtually impenetrable.

Our skin is also covered with hairs, while our mucous membranes and the surface of our airways have tiny cilia, which all wave in the same direction to usher back out any incoming bacteria. That cilia layer is what smokers destroy in their airways, immediately putting part of their defences out of action.

Alongside your skin and mucous membranes, healthy flora also form part of your frontline defences. Healthy bacteria make it clear to would-be intruders that they're unwelcome by being present in just the right amounts, and by fermenting sugars in your body, creating a slightly more acidic environment, which wards off bad bacteria.

The next element of defence is your bodily fluids. All bodily fluids flow in the necessary direction to remove microbes via the shortest route. When we blow our nose or cough, we're propelling fluid out. Diarrhoea and vomiting are also fantastic mechanisms of defence against intruders.

If there's something in your digestive tract that shouldn't be there, your body reacts immediately to get the garbage out. Other methods: eyes watering when a speck of dirt gets in them, saliva production, oil from our sebaceous glands, bile produced by the liver to kill bacteria in the gut, and stomach acid, which is anything but bacteria-friendly. Your front line of defence really is a robust castle!

But no fortress is impenetrable, and sometimes pathogens manage to break through the frontline defences. That's when the second line of defence is deployed: your white blood cells.

The White Army

Blood isn't just blood: it's made up of several different kinds of cell. Red blood cells have little involvement in the body's defence system — their main job is to transport oxygen around. Platelets enable the blood to clot and to form scabs over wounds. And then there are the white blood cells, or leukocytes, which are the soldiers of your body's defence system! Just like soldiers in an army, white blood cells have different ranks and different functions, and each type of white blood cell has its own weapons.

Among the soldiers in this army are the natural killer cells and the phagocytes. This part of your defences is known medically as the innate immune system, as it is active from birth. Even without knowing what pathogen is involved, these cells know what's expected of them, which explains this system's other medical name: non-specific immunity. It attacks all pathogenic microbes in the same way.

The natural killer cells can be considered the army's snipers. On contact with an intruder, they begin firing out their perforins, which are like their bullets. As the name suggests, these perforate the offending bacteria or nascent tumour cells to destroy them.

Phagocytes, which I referred to as Pac-Man cells in the Introduction to this book, have a different job to do. They are the eaters. Phagocytes can smell intruders in the bloodstream, detecting the particular chemical substances that bacteria produce in the body; the scientific name for this process is chemotaxis. The eaters then charge the offending bacterium, fold themselves around it, and ingest it, thus destroying it. At this stage, a substance called interleukin is released, which enables the white blood cells to communicate with each other. Interleukins travel through the body to the brain, where they raise the alarm and turn up the body's thermostat, which results in fever. By increasing its temperature in this way, the body is better able to keep the invading bacteria in check: an increased heart rate means the white blood cells are pumped more quickly through the circulatory system, rapidly mobilising the entire army.

The ranks of the phagocytes are made up of many types. So far, I've lumped all of them together, but let me now be more specific and differentiate between the different kinds of these little Pac-Mans. Neutrophils are the most abundant type of white blood cells. They gobble up pathogens, but then give up the ghost themselves. Actually, you're already very familiar with neutrophils: they're the cells that turn into a load of gunk after they've feasted on bacteria. That gunk is the fluid we know as ... yes, pus, that's right!

A second type of phagocyte is the macrophages. They have the ability to leave the bloodstream to patrol other tissue, on the lookout for unwelcome guests. Anything that looks suspicious has to face the music. Upon locating a pathogen, a macrophage will wrap its tentacles around the intruder, suck it in, and destroy it, then spit out the leftovers. They can carry out this action repeatedly.

Eosinophils spring into action when there's a parasitic infection or an allergic reaction in the offing, working together with mast cells, which are the body's producers of histamines. Also noteworthy are the dendritic cells — they can easily wrap their tentacles around any intruder, attacking everything and passing the resulting material on to other cells. They present pathogens to other immune cells on a silver platter, so to speak.

All these cells in your body are able to communicate with each other. If a cell in your nose is attacked by a virus, it alerts neighbouring cells by releasing interferon, which functions as a kind of red flag, letting the other cells know they need to protect themselves. In the meantime, a whole load of other proteins begin circulating in your blood, preparing your immune system even further for rapid activation of all these processes. This chain reaction is called the complement system. It supports all the processes undergone by cells, and activates your body's inflammatory response, causing it to produce substances that provoke an inflammation, which is necessary for you to recover.

Inflammation

Inflammation is something we need?

Absolutely! Inflammation causes redness, heat, swelling, and pain. I like to use the example of a mosquito bite, which triggers a kind of mini-inflammation: you get some redness and swelling in a very localised area, the bite feels warm, and it hurts. That's your body's inflammatory response: your white blood cells dilate your blood vessels by releasing pro-inflammatory substances such as histamines into your blood. This gives the white blood cells more room to circulate. At the same time, prostaglandins are sent to your brain, which increases your body temperature even further.

This raises an important question: when you have such symptoms, should you take anti-inflammatory or antipyretic (fever-reducing) drugs to suppress them? Thinking about it logically, the answer is: you should take the smallest amount possible. As well as suppressing your symptoms, they also suppress your immune system and reduce the effectiveness of your body's defences. Of course, there are cases where this is necessary — for instance, if your temperature gets too high or you're in too much pain. But, strictly speaking, by taking medication to reduce an inflammation, you are actually working against your own body.

So far, so good. We know that an intruder has to overcome significant physical barriers to enter our well-defended system. And if it manages to get through nonetheless, your white blood cells muster and prepare for attack. Now comes the third line of defence: the adaptive, or acquired, immune response. Who are the main protagonists on this battleground? Yet another kind of white blood cell. Reinforcements are on the way for the Pac-Man battalions — in the form of lymphocytes.

The Strategy of Your White Army

Are you still following? Are you wondering what exactly happens, and when? Who does what and at which point in the process? Front line, second line, third line ... and a load of unfamiliar words. I'm sorry. Let me summarise it for you again: the identity cards below show you the features of the various blood cells. I will then explain it again using a simple example.

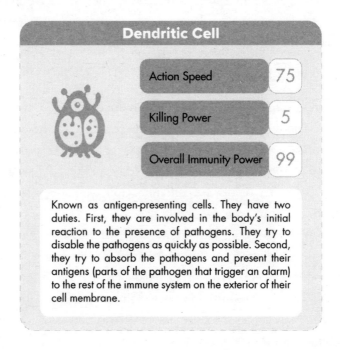

Dendritic Cell

Action Speed	75
Killing Power	5
Overall Immunity Power	99

Known as antigen-presenting cells. They have two duties. First, they are involved in the body's initial reaction to the presence of pathogens. They try to disable the pathogens as quickly as possible. Second, they try to absorb the pathogens and present their antigens (parts of the pathogen that trigger an alarm) to the rest of the immune system on the exterior of their cell membrane.

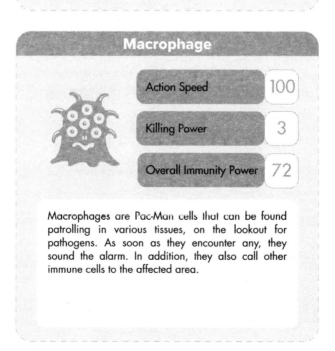

Neutrophil

Action Speed	95
Killing Power	60
Overall Immunity Power	48

An essential subdivision of the immune system. These are short-lived Pac-Man cells that take part in the initial response to an infection. Neutrophils phagocytise (eat) microorganisms to destroy them, but they die themselves in the process and become pus. A deficiency in neutrophils leaves you more vulnerable to bacterial infections.

Macrophage

Action Speed	100
Killing Power	3
Overall Immunity Power	72

Macrophages are Pac-Man cells that can be found patrolling in various tissues, on the lookout for pathogens. As soon as they encounter any, they sound the alarm. In addition, they also call other immune cells to the affected area.

Eosinophil

Action Speed	80
Killing Power	45
Overall Immunity Power	40

Eosinophils are innate immune cells containing relatively big, regularly shaped granules that are bright orange in colour. The most important function of eosinophils is to destroy parasites. They also play a part in allergic reactions.

Natural Killer Cell

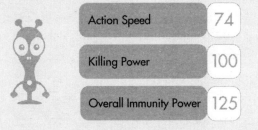

Action Speed	74
Killing Power	100
Overall Immunity Power	125

Natural killer cells have built-in receptors for immediately recognising pathogens. They target infected cells and kill them. Natural killer cells also play an important part in identifying and removing damaged and malignant cells before they can develop into tumours.

Suppose you trip while walking down the street and graze your knee. In one fell swoop, your entire front line of defence is gone. The unassailable fortress is now open to attack. Pathogenic germs rush in unhindered. Your body immediately raises the alarm in your brain, before the microbes can penetrate too deeply. In other words, your inflammatory response is triggered.

Mast cells in your injured knee release histamines, which dilate the blood vessels surrounding the graze so that blood can flow to it more easily and bacteria are flushed away. This process is called vasodilation and is responsible for the slight swelling and redness of your grazed knee. It also increases the blood supply, which means more cells can be delivered to the site of the injury.

The histamines also send out signals to the phagocytes, which mobilise immediately. The neutrophils are the first to arrive at the scene of the accident. They start cleaning up like crazy, gobbling up bacteria, but, in doing so, they also commit mass suicide — you remember, these little Pac-Man cells stage a valiant defence, but don't survive the battle. Pus develops, and the macrophages rush to their comrades' aid. They clear up the waste and the mess that the neutrophils have made. The traffic in your bloodstream gets busier, and the waste is transported away to your body's sewage plant: the lymphatic system. Here, the waste products gather in your lymph nodes and are then excreted via your blood.

If the infection is serious enough, the local troops can still eventually be overpowered. If the sheer number of invaders is too great, the phagocytes sound the alarm by sending chemical substances to your brain, which turns up your body's thermostat. The resulting fever speeds up the cells' metabolism, enabling them to recover more quickly. At the same time, your liver and spleen receive the important instruction to stop releasing zinc and iron into the bloodstream for a while. After all, bacteria feed on iron and zinc, making them more difficult to vanquish. In this book's chapter on medicines, you'll read that it's best not to treat an iron deficiency with supplements during an infection: this is the reason.

But sometimes even these actions aren't enough. Then it's time to call in the special forces. Time for the third line of defence. Time for the main players in this story, the B and T lymphocytes.

Know Thine Enemy

What's true of real war is also true of the immune system: if you want to win, it's important to know your enemy. The process I described above is part of your innate immune response. Its effect is fairly universal and based on a zero-tolerance policy: all intruders are exterminated. But you have another immune system, one which is not with you from birth. It develops gradually as you grow up. It's also far more selective in its modus operandi. Before it can do anything, it has to be taught what needs to be done. It has to learn to see bacteria and viruses as threats before it knows to attack them.

One form this education takes is through contact with 'dirty' things, such as, for example, door handles, toilet brushes, someone near you sneezing. Other ways your acquired immune system gets smarter is through vaccines and previous infections. When your third line of defence encounters such a threat for the first time, the experience is stored for future use and never forgotten: there are special cells in your body that keep these memories for you. This memory bank function is the most important way the acquired immune system differs from the innate immune system.

There's Humour in Your Body

The first part of the acquired immune system is humoural immunity. In this sense, the word *humour* has nothing to do with whether you appreciate a good joke, but comes from the Latin word meaning 'body fluid'.

According to the medical theory of the Middle Ages, a person's health and temperament were determined by four bodily fluids, known has the four humours: phlegm, blood, yellow bile, and black bile. Depending on which fluid was predominant, patients were deemed hot-headed or full-blooded, bilious or melancholy, phlegmatic, and so on. Naturally, there were different treatments for each humorous state, including things like bloodletting. If all four humours were in perfect balance, the patient was healthy.

Specialised white blood cells called B lymphocytes arise and grow from stem cells in your bone marrow, which is where they learn to distinguish friend from foe so that they don't start attacking your body's own cells. Once they've mastered that, they're allowed to roam freely. They screen your blood and lymphatic fluid, where they fight viruses, bacteria, and other undesirables. They're the ones who produce those all-important proteins — antibodies.

Every potential pathogen has antigens, certain prominent molecules on the outside of its cell wall. These can take various forms, and may be triangular, rectangular, or round in shape. B lymphocytes can produce antibodies to catch a pathogen by its antigen. The antibody has a location where it can bind to the pathogen, which is the precise negative of its prey's surface molecule in shape, like pieces of a jigsaw puzzle. So we have to make a specific antibody for each antigen. Antigens are found on viruses, bacteria, fungi, and even the cells of your own body if they are diseased (such as when they become cancerous).

The B lymphocytes from your humoural immune system are trained to recognise different antigens. Some may be trained to recognise the coronavirus, for example, while others specialise in the flu virus, and so they can identify which pathogenic germ has invaded your body. They are assisted in this by the natural killer and dendritic cells, which present antigens to the B cells. Do you remember those innate cells? The main players in our second line of defence, they eat up or shoot down everything they come across. Then they strut around showing off their spoils of war, making the body parts of the enemy easily identifiable for our B lymphocytes.

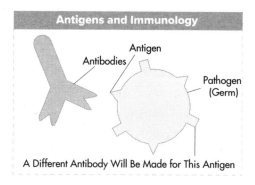

A Different Antibody Will Be Made for This Antigen

Suppose for a moment that you've contracted measles. The virus enters your body and looks around for a nice place to take off its protein coat. If this is the first time you've come into contact with the measles virus — i.e. if you've never had measles before and haven't been vaccinated against it — your B cells won't immediately know what to do. But since they are kindly offered the antigen through the actions of your second-line defences, they can get to work. A zealous B cell examines the antigen, builds the right antibody to fit it, and calls on its comrades to come and help. The other B cells arrive en masse as reinforcements. Some turn into antibody factories, known scientifically as effector B cells. Others learn and memorise the recipe for making the antibody; these are called memory B cells. If you've already had measles or are vaccinated against the disease, you will already have memory B cells with the right knowledge circulating in your blood when you encounter the virus, and the production of antibodies will kick in all the more quickly. When a measles virus comes across a B cell before it's had a chance to settle somewhere, the B cell immediately whips out the big guns and binds its antibody securely to the measles virus's antigen. And, again, it immediately calls in reinforcements: the B cell duplicates itself at lightning speed, rapidly creating an entire army of B cells with specific antibodies against measles.

So, there are two categories of B cells: effector cells, which are deployed during combat, and memory cells. They hang around in your body for the rest of your life, always ready to defend you, should you be attacked again by a pathogen they know.

Effector cells can produce up to 2,000 antibodies per second. Those antibodies then drift freely through your system, marking every virus antigen they encounter with a kind of label, a red flag to indicate that those cells must be dealt with — by others, as they can't do it themselves. The antibodies-with-antigens created in this way stick together — a process called agglutination — to form clusters, which are then cleared up by the Pac-Man cells of your second line of defence. This is an example of the brilliant collaboration between your innate and acquired immune systems, keeping you as healthy as possible in the short term, while also protecting you in the long term with memory cells.

This process occurs not only when you get infected with measles, but also when you catch the flu, or any other infection for that matter. It can also be stimulated artificially by vaccination: vaccines contain attenuated (weakened) or facsimile antigens, which don't make you sick but do provoke an immune response. No red flags are raised, but memory cells are created nonetheless. For measles, the resulting immunity is basically lifelong. In the case of other infections, the antigen is changing slightly all the time, which fools your B cells and means your body has to build up fresh immunity all the time. That's why you can get the flu more than once: the memory cells created in response to the bout you had last year can't recognise the viruses responsible for this year's flu season.

Fake News

In some cases, temporary immunity can be 'forced', such as when patients at high risk need to be protected against a certain disease during hospitalisation. Studies have already been carried out in which blood from COVID-positive patients was given to sick high-risk patients to provide their system with some antibodies for a short time. Such protection is only temporary, because the patient's B cells aren't activated. Sooner or later, patients have to go through the whole cycle to make real memory and effector cells. Until they have those, it's a temporary case of fake news.

Pregnant women also pass on passive antibodies to their babies via the placenta. Even after birth, new babies have passive protection. They receive temporary antibodies via their mother's milk. Thus, babies are protected despite their bodies having almost no immunity of their own. That lack of independent immunity explains why newborn babies often get sick or feverish. The acquired immune system constantly has to learn new things, and getting sick is the quickest way of teaching it.

Plan D

Let's briefly go back to the beginning: the first line of defence has been broken and the second appears to be too weak, so it's time to activate the humoural immune system, with its B lymphocytes. Antibodies are produced and everything is flagged and neutralised. But B cells are too peace-loving, and viruses and other pathogens remain unkilled. So, when even those special forces are unable to outsmart the enemy, it's time for Plan D!

And Plan D, dear reader, is the second division of our third line of defence: cellular immunity. The fight no longer takes place on the main battlefield; now the gloves are off, and it's cell versus cell. The main players in this show are the T lymphocytes.

T cells are multitaskers: they activate inflammatory responses, free body cells that have been taken hostage by viruses and bacteria, call in the macrophages, and recruit other T cells ... Actually, to some extent, they are the immune system's annoying busybodies, but they are extremely important members of the team. Not only do they regulate all kinds of processes, they can also take the reins themselves, if necessary, when no one else can solve the problem.

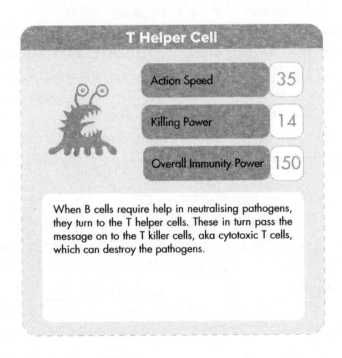

T Helper Cell

Action Speed	35
Killing Power	14
Overall Immunity Power	150

When B cells require help in neutralising pathogens, they turn to the T helper cells. These in turn pass the message on to the T killer cells, aka cytotoxic T cells, which can destroy the pathogens.

T Killer Cell

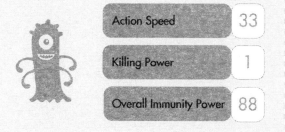

Action Speed	30
Killing Power	98
Overall Immunity Power	129

T killer cells require the aid of T helper cells to become completely active in recognising an antigen. T killer cells shoot other cells with perforins. The T killer cells survive and can go on to eliminate multiple targets.

T Regulator Cell

Action Speed	33
Killing Power	1
Overall Immunity Power	88

These cells prevent the immune system from reacting to harmless molecules (mostly our own tissue) and help to calm the system back down after it has eliminated pathogens. However, tumours can use these cells to protect themselves from attack by the immune system.

Collateral Damage

There are three kinds of T cells: helpers, killers, and regulators. T helper cells can't kill, but they can communicate with B cells. When the B cells become so busy producing and distributing antibodies that they can no longer cope, they call in the T helper cells, which pass the message on to the T killer cells, also known as cytotoxic (cell-killing) T cells, which can kill the pathogens.

The pathogens' identifying surface molecules, their antigens, are presented to the T killer cells by our second-line phagocytes. The antigens are stuck on an MHC protein on the phagocyte's cell membrane. The protein's full name — which you can use to impress your friends — is major histocompatibility complex. You could also casually drop the term in your next Zoom meeting, to wake everybody up when their attention is waning.

Other cells have MHC proteins, too. In fact, every cell in your body has MHC proteins on its membrane: they let your immune cells know which cells are healthy and which are sick, which are foreign agents, and which are cells from your own body. You can view the MHC system as a locking device for every cell. T cells can't do anything to other cells unless their key fits the lock. This might make you think that T cells are designed to perfection, but that isn't the case. The lock on the cell makes it clear to your body that the cell is okay, so immune cells stay away from it. If a cell becomes cancerous, the MHC lock changes and your immune cells know they can now destroy it. Macrophages and B cells only have one key, so they only work on one very specific type of cell. Unfortunately, our T lymphocytes aren't all that smart, so our Plan D heroes aren't always able to distinguish good cells from bad, and they drag many of the body's own cells along to their doom. The T killer cells put all intruders up against the wall and shoot them.

This process is the basis for understanding how immunity can conquer but also cause disease. In such cases, it's said that the patient's autoimmunity goes completely haywire, and her body busily starts destroying its own tissue. Examples of T-helper dysfunction, in which the communication between the humoural and the cellular response completely breaks down, include thyroid disease, lupus, type 1 diabetes, multiple sclerosis, psoriasis, rheumatoid arthritis ...

The reason why T and B cells suddenly turn against a patient's own body is still not understood in detail. The MHC lock-and-key system certainly plays an important part. The T busybodies also bear some responsibility, as they are supposed to tell their B and T friends to cool it. Genetic and environmental factors are certainly involved in some way, too. Certain viruses can disrupt the immune system to such an extent that initial immunity turns into autoimmunity in the long term.

Besides painkillers and anti-inflammatory drugs, the stress hormone cortisone is often prescribed for such conditions, because it suppresses the immune system. One of the drawbacks of such a treatment is that it leaves the patient's body more vulnerable to external invaders. However, we are getting a much better picture of each condition all the time, so treatment is becoming increasingly more targeted. For instance, immunotherapy can suppress certain T cells, and a great future is predicted for the use of stem-cell therapies.

Hypersensitivity

Our immunity is a fantastic system, when it works properly. When it isn't working as it should, when it is overly sensitive, your immune system can provoke an autoimmune reaction or an allergic reaction. Such reactions can sometimes cause discomfort and are otherwise harmless, but they can also be fatal.

One type of hypersensitivity is called anaphylaxis. Allergens that are known to cause this reaction include certain foods, such as tree nuts, shellfish, eggs, peanuts, and seafood, but also the bites or stings of certain insects, and some medicines such as penicillin or salicylic acid.

So, what, precisely, happens in such cases? For some mysterious reason, a B lymphocyte begins producing a special antibody that pushes another of the immune system's cells, the mast cells, very hard to release histamine. When the reaction is local, only a specific area goes red or swells up, like a mosquito bite or a wasp sting. But the reaction can also be systemic, when all of the body's mast cells suddenly receive the instruction to release histamine: *all* your blood vessels become dilated, your blood pressure plummets, and you collapse — anaphylactic shock! Administering adrenaline (epinephrine) very quickly is the only way to get those blood vessels to constrict again.

Another type of hypersensitivity usually happens some-what more slowly and has to do with a different kind of antibody. Phagocytes play a major part in this reaction. The best example of this kind of hypersensitivity is erythroblastosis fetalis, or Rhesus-factor incompatibility. If during birth the blood of a mother who is Rhesus negative comes into contact with the blood of a baby who is Rhesus positive, the mother's body produces antibodies that will attack the blood cells of the baby in her next pregnancy. If nothing is done, then that next baby will die. Nowadays, mothers in that situation are vaccinated to prevent them producing the dangerous antibodies.

A third type of hypersensitivity is the formation of immune complexes. Antibodies bind to antigens, clump together, and sometimes accumulate in huge complexes, which float around your body. Some types of food can trigger the formation of immune complexes, creating those floating clusters, which start to overstimulate the immune system elsewhere (in the thyroid gland and the joints). This is how food hypersensitivities arise. Autoimmune disorders such as lupus or rheumatism are caused in a similar way. Your body suddenly interprets certain cells as foreign and that sets a whole cascade of reactions in motion ...

A fourth type is delayed hypersensitivity — not caused by antibodies but by body cells. A nickel allergy, for instance, can lead to this. If you wear an earring that contains nickel, you may develop an immediate reaction: that's caused by your mast cells producing histamine, and, if it only causes skin irritation, this kind of allergy is harmless. All you need to do is simply take out the offending earring and you're cured. But in some cases, it can progress to a delayed reaction, which only appears 48 to 72 hours after contact with the allergenic substance. If this hypersensitivity becomes systemic, it can cause extreme cases of eczema or fibrosis all over the body.

'At first they'll ask you
why you're doing it.
Later, they'll ask you
how you did it'

Pankaj Yadav

LIFESTYLE

Congratulations! You have now reached the most important part of the book. This ties in nicely with the question that increasingly occupies all of us: can you improve your immune system by changing your lifestyle? That's the Big Question I'll be answering for you in this chapter. I'm more than happy to quote Arthur C. Clarke here:

> *Every revolutionary idea seems to evoke three stages of reaction. They may be summed up by the phrases:*
>
> *1. It's completely impossible.*
> *2. It's possible, but it's not worth doing.*
> *3. I said it was a good idea all along!*

The life cycle of every innovation, whether it's in business or healthcare or elsewhere, follows a certain pattern. It begins with a small group of innovators who try out the innovation. They then influence the second group, the extremely progressive 'early adopters', who help to spread the word. Next, a sizeable number of people jump on the bandwagon, and they're known as the 'early majority' — and, eventually, everyone is on board. Lifestyle and behavioural changes spread according to the same pattern.

What effect does your lifestyle have on you? What are the reasons for adopting or adapting a certain lifestyle? We live in a world where chronic illnesses are on the rise, and they are a great burden on our healthcare systems. Our primary goal is to stop such chronic diseases from getting worse, but also to prevent them from developing in the first place, and, if possible, to reverse them when they do. I'm convinced that we can achieve this, with the right kind of 'lifestyle medicine'!

As an introduction to this important chapter, I'd like to look briefly at biology. How can sensible behaviour when it comes to eating, exercising, sleep, and stress — to name just a few areas — ensure that we can prevent, stop, and reverse this heavy burden of chronic illness? There are many factors that influence our bodies, and I'm going to take a closer look at them now.

Trust Me, I'm a Doctor

I've been trained to think in terms of disease. I examine a patient's symptoms, pin a diagnosis on them, and prescribe treatment. After all, that's what I was trained to do. Is this method foolproof? Well ... What doctors do is perform an algorithmic probability calculation: we match the symptoms to a diagnosis and are 95, 96, or 99 per cent sure what a complaint could be. We make use of technical aids, like medical imaging technology or blood tests, which helps us refine the diagnosis and decide how a patient's symptoms can best be treated.

This method works perfectly well if the aim is to cure ailments, but, when it comes to preventing disease, I feel it is quite limiting. You draw a direct line connecting a symptom to a treatment. In both sports medicine and preventive medicine, it's at least as important to follow the line back in the opposite direction: Why does a person develop certain symptoms? Is there perhaps something wrong with their energy metabolism? Do they have a genetic defect, or a problem with inflammation or their immune system? There's usually a biological cause behind the artificial divisions between symptom, disease, and treatment. And, as a doctor, I'm increasingly interested in that biological cause!

We doctors aren't trained in keeping people healthy; we're trained to treat people who are already sick. That's a good thing, and I'm not saying we should stop doing it, as people are always going to get sick sometimes. But we should receive at least as solid a grounding in how to keep people healthy, through screening for medical risk factors, proposing changes in lifestyle and nutrition habits, and so on. In this way, we can optimise you, and help you become the best possible version of yourself.

This is what good lifestyle medicine is all about. Not waiting until the symptoms of diabetes appear before intervening, but anticipating them before they manifest themselves. Not waiting until the first heart attack to discover a patient's arteries are blocked and an operation or powerful medication are the only options.

Chronic conditions can't be treated in the same way as acute illnesses. It's not enough to treat diabetes with drugs that lower blood-sugar levels. That's just like sweeping the dirt under the carpet. We need to look for the underlying mechanisms.

The Four P's

Even small changes to your lifestyle can bring great benefits. Personal lifestyle programs like this are the forerunners of personalised medical care. The medical care of the future is also sometimes called *P4 medicine* —

P for preventive: we will organise ourselves in such a way as to prevent certain conditions, which will also reduce the financial burden on our national healthcare systems.

P for predictive: just as banks and corporate marketing departments do, we will gather data and carry out DNA tests, to find out which people face which risks, so that we can anticipate and predict medical data.

P for personalised: we will tailor medical care to the patient in a very personalised way. We will be aware of what kind of advice works for you specifically. There's no point in making you listen to all kinds of experts and academics telling you what 'healthy eating' really means. What's good for you might not be good for your neighbour. Lifestyle guidelines can't be reduced to generic advice; they need to be linked uniquely to your personality. The guidance you receive won't necessarily be the same as that given to your children, your partner, or your friends.

P for participatory: we are evolving towards a platform-centred society. Just as banking now mainly takes place on online platforms, health and lifestyle will also become increasingly 'platformed'. This will make it easier for doctors to monitor and track their patients. You share your data on a platform and the system can process it, personalise the medical advice you receive, and create predictive models for you. The more data the system has, the smarter and more accurate it will become. This is known as a learning healthcare system.

I really can't wait to see the healthcare of the future!

The Best Time to Plant a Tree Is Ten Years Ago

In this chapter, I'm going zoom in on the mechanisms that underlie chronic illness. Those mechanisms depend greatly on the lifestyle choices we make. But it's not a one-to-one correspondence: any given lifestyle change will not only have an effect on your risk of developing diabetes, for example, but will also affect your immune system, your risk of cardiovascular disease, your weight ... Lifestyle and behaviour form a network! The number of people consciously choosing to pursue a healthier life is growing all the time. The best way to make that change is to take baby steps rather than trying to go from zero to 100 in two seconds flat. The best time to plant a tree is ten years ago, but the next best time is now. In other words, it's never too late to change your habits. Lifestyle is a fantastic weapon against inflammation, against a failing immune system.

Right, that's enough philosophising. It's time to get down to brass tacks! There are a number of factors that influence our physical state and the optimum operation of our immune system. I will go through them one by one, using a number of fictional characters as illustrations. Any similarity to real persons living or dead is purely coincidental, and no animals were harmed in the writing of this book.

Nutrition

Servaas (purely coincidentally, a man of 39) is currently relatively healthy and follows a balanced diet according to the 80/20 principle. But, to be perfectly honest, healthy eating has been a struggle for him for a long time. (And now I will switch to using the pronoun *I*, as I'm talking about myself, of course.)

I've read and tried to apply many theories, but for the longest time never succeeded in eating healthily or maintaining a healthy body weight. I worked in top-level sport and saw what a great impact nutrition has, but I was rather late in realising the importance of a healthy diet for myself. The more books I read on the subject of exercise, the less exercise I took ...

And the bathroom scales didn't lie. I was low on energy and generally didn't feel good. But that's the great thing about being human: although we often see bad feelings as something negative, they're just our body's way of sending us a message. Do something, dig yourself out of that hole. It's an opportunity to grow as a person!

What it comes down to is this: each of us is a biological system, and we are designed to eat organic things. Anything that is too processed or too refined is not good for us. Our body can get all the carbohydrates, proteins, and calories it needs from a plant-based diet. Just think of all the different vegetables and fruit and grains and nuts and seeds and beans and pulses there are! What you *do* eat is far more important than what you *don't* eat.

First of all, you need to make a deal with yourself and agree on a nutritional concept: what are you going to eat, when are you going to eat it, and how much? I'm going to help you create a nutritional concept for yourself.

You Are What, When, and How You Eat

The first thing to talk about when it comes to what you eat and what you don't is dietary restriction, which is the first part of our nutritional concept. What should you leave out of your diet, and why? You might choose a low-carb diet if you want to lose weight. Or you might opt for a vegetarian diet, or a diet adapted to your food intolerances ... There are many options. Also impose a time restriction on yourself: only eat during a certain time window and give your body some rest at other times. Intermittent fasting, using the safe and effective 16/8 method, fits this time frame perfectly. You can also set yourself a limit on how much you're going to eat, which is known as calorie restriction.

I personally benefit most from a combination of these three elements. I eat very few carbohydrates, especially in the evenings. I do eat vegetables and fruit, but I avoid all other kinds of carbs. I follow the 16/8 eating pattern (fasting for 16 hours of the day, leaving the other eight for eating — e.g. between 11.00 am and 7.00 pm), which is my time restriction. I also unconsciously impose a calorie restriction on myself by limiting my eating window and by avoiding mindlessly grazing on empty calories in the form of cookies and other junk food. Sometimes I manage to keep to it really well; sometimes I don't. But when I relapse, I can always fall back on my nutrition concept.

Decide for yourself what fits in best with your daily routine, what's feasible, and what isn't. Is your focus on losing weight, or are you trying to become as healthy as possible? Do you want to perform better in a certain sport, or do you want to optimise the functioning of your immune system? Once you've set your goal, you can adapt your eating pattern correspondingly.

Nothing Is Set in Stone

I often see patients who fast intermittently, just as they're told they should, but still don't lose weight. If that's you, then it's time to get personal with your dietary advice. Remember the four P's? You and your doctor need to check whether there are any medical reasons for your lack of weight loss. Possible reasons are a hormonal disorder or a problem with your microbiome. Or perhaps your own immune system is preventing you from losing weight for one reason or another. A good example of this is when inflammation, the foundation of immunity, becomes chronic inflammation. In such cases, the first thing to do is to solve the underlying problem.

A dietary plan tailored to your needs will help you sleep better, give you more energy, reduce the cravings you have throughout the day, and improve gastrointestinal issues like reflux, heartburn, and bloating. However, if these improvements fail to materialise, then it's time for a different strategy. I am and always will be a doctor, so my recommendation is a thorough medical examination. After all, my motto has always been: measure to manage!

Can you just go to your doctor and ask 'to have your immune system analysed with blood tests'? Not quite. But you can have certain blood tests that will give you information about blood levels and biomarkers that might indicate what the problem is.

Let's begin with vitamin D. An ideal blood level for this vitamin is around 60 nmol/L, although most laboratory reference values state that 30 is fine. Our body requires sunlight to be able to synthesise this vitamin. Since the level of UV light received is usually too low for many people, we also have to rely on our diet as a source of vitamin D. We get it from foods such as oily fish, real butter, or eggs. One absolute vitamin-D booster is something older readers might remember: that daily spoonful of cod-liver oil before bedtime ...

You can also make up for a deficiency in vitamin D by taking supplements. Vitamin D is fat-soluble, so you must always take such supplements with a high-fat meal, otherwise it will not be absorbed as well by your body.

Vitamin D is more than vital for your body. It plays an important part in staving off cardiovascular conditions, autoimmune diseases, and cancer. It also increases the body's ability to absorb calcium, which we need not only to make our bones stronger, but also for our nerves and blood vessels. Muscles use up a great deal of calcium when they contract. During our crash course in immunology, I told you how white blood cells exchange substances to communicate with each other. Communication between T helper and T regulator cells is based on calcium. So, a calcium deficiency can disrupt the functioning of your T cells. Calcium is found in dairy products, but also in broccoli and various types of cabbage, oily fish such as mackerel, salmon, and sardines, and whole grains. But note also that calcium is the substance that sneaks into your artery walls, where it can cause arteriosclerosis — which only goes to show that indiscriminately popping as many supplement pills as possible is not the answer.

You can also measure the levels of fatty acids in your body — more precisely, your omega-3 levels. Why is that so important? Every cell in your body actually consists of fluid surrounded by a lipid layer. Each cell membrane is made up of fatty acids, and the same goes for immune cells, all of which need to be in perfect condition if they are to perform their specific functions as well as possible.

Among the fatty acids there are two special families: omega-3 and omega-6. The omega-3s have two main benefits. The first is that they make our blood platelets less sticky, reducing our risk of blood clots. One of the first studies of omega-3 fatty acids was carried out in the 1970s. It examined the effect of salmon oil on the blood-cholesterol levels in a group of test subjects. After 28 days, the researchers discovered, the group had much lower levels of triglycerides — fats that can contribute to the clogging of the arteries. Omega-3 is also involved in the immune system, and this brings us to its second set of benefits. It helps determine whether a T cell becomes a helper or a regulator, supports the phagocytic activity of macrophages, and has a major anti-inflammatory effect. When the immune system's second- and third-line defences kick in, cytokines (proteins that provoke an inflammatory response in the body) are released. This is counteracted by omega-3, preventing the reaction from tipping over into a massive inflammatory overreaction. Recent research on animals with severe autoimmune diseases shows that giving them high doses of omega-3 has promising results ... To be continued!

Macronutrients

There are many possible reasons why someone's immune system might not be functioning properly. In some countries, undernourishment is the cause of many problems, especially in combination with poor sanitary conditions. Undernourishment results from a lack of sufficient macro- and micronutrients. I'll go into more detail on micronutrients in the chapter called 'Inside the Medicine Cabinet', but let's now take a brief look at macronutrients.

In countries like ours, macronutrient undernourishment is rare (except in people with eating disorders). What we have to contend with, on the other hand, is macro-*overnutrition*. Obese people, for example, often also suffer from a poorly functioning immune system.

There are four main groups of macronutrients:

- Proteins are pretty important for the immune system because they ensure that the body's cells can divide and produce new cells. Both the older population and very young children need to take in the correction proportions of protein to fulfil the needs of their immune system and their body as a whole. But too high an intake of protein can lead to health risks, too. Prolonged overconsumption of protein is associated with bone-growth and repair problems, kidney and liver dysfunction, and an increased risk of both cancer and vascular disease. Current guidelines recommend 0.8 grams of protein per kilogram of body weight per day for adults. The recommendation for adolescents is one gram per kilo of body weight per day. For growing children, the recommended amount is 1.5 grams.
- Carbohydrates are the next group. Non-natural carbohydrates, especially refined sugars, paralyse those Pac-Man cells I talked about in great detail earlier, and this effect can last for a considerable period of time: phagocytic activity sinks as our insulin levels rise. Ingesting sugar also temporarily reduces the amount of vitamin C your body can absorb — and that vitamin is an

extremely important nutrient, as it's the energy provider to our white blood cells.

- Fats are the third group of macronutrients. As I explained earlier, fatty acids are very important because they form the membrane of every cell in the body. Experts recommend that fats should make up 20 to 35 per cent of your daily food intake. Based on a diet of 2,000 calories, that works out at 44 to 77 grams of fat per day. Just read the labels on your food to keep track of the amount of fat you're eating each day. Fat is an important source of energy and provides the body with some essential fatty acids that it can't produce itself. We need fat to be able to absorb vitamins A, D, E, and K. But be careful: not all fats are the same. The best kind are the unsaturated fatty acids, which are liquid at room temperature. Saturated, or hard, fats are far less interesting, and trans fats are best avoided altogether. I could write an entire book on fats alone, but that isn't my intention here.

- Fibre. Dietary fibre actually consists of carbohydrates, but this time of the unrefined kind. From this we know that they are bound together in complex ways and can't be broken down rapidly into smaller units. They perform a number of important functions for the immune system, and stimulate the cells around the GALT — the gut-associated lymphoid tissue. The GALT is a system of lymphatic tracts and groups of cells around the gut that function as the vanguards of the immune system. They are in our gut, but our tonsils are also part of this system. The gut can also benefit indirectly from dietary fibre, because it provides the right breeding ground for good bacteria to grow, which can then help in the fight against bad bacteria.

Alcohol

I can't really call it a macronutrient, but I want to deal briefly with alcohol here, and, more specifically, with the harmful effect it has on the immune system. When a virus enters your body, your immune system springs into action. The healthier you are, the healthier your defences are, too, and the faster you can fight off the virus and recover from the symptoms. Alcohol, however, impairs the quality of the immune cells in several important organs of your body — in your lungs, for example, where the cilia are also destroyed by alcohol consumption; the cilia are part of your frontline defences, which are thus weakened in their fight against pathogens. Alcohol also fosters inflammation. Drinking alcohol has the effect of disinfecting your gut, but that's the last place that needs sweeping clean of microbes. On the contrary, your gut bacteria are extremely important for an optimally healthy immune system.

Alcohol has another detrimental effect: it distracts the body from its usual activities. It negatively affects the quality of your sleep and the operation of your metabolism. Your body has to detoxify and break down the alcohol, when it should be busy breaking down the waste products that normally accumulate in your system. That places an extra burden on your body.

How much is too much, you may be asking. That differs greatly from on individual to another, but the bottom line is that you shouldn't drink every day. Skip a week every now and then, and remember that alcohol does more harm than good.

The Microbiome

I've mentioned it over and over already, so now it's time to take a closer look at your microbiome.

Our gut is completely sterile until we are born. From that moment on, our personal microbiome begins to form. First, from bacteria in the vagina during birth itself — nowadays, babies born by caesarean section are deliberately swabbed with vaginal bacteria to seed their microbiome — and, later, from routine exposure as we grow up, starting with that first mouthful of soil no child can resist.

Our gut performs a very important function. It's the body's largest surface of contact with the outside world — it's big enough to cover a couple of tennis courts! The immune system, and certainly its first attackers, often act in the area of the gut. That's why it's so important to know your gut microbiome well.

Gut disorders can be divided into three categories:

- Abnormalities in the anatomical structure of the gut, such as diverticulitis, in which a pouch (or multiple pouches) forms in the gut wall, where remnants of food or faeces can get stuck and cause inflammation. Polyps can also occur and are sometimes a precursor to bowel cancer. This category also includes haemorrhoids, which occur when a blood vessel in the gut becomes swollen and gets in the way.
- Functional gut disorders, such as irritable bowel syndrome, which causes abnormal bowel movements, bloating, and digestive problems. A poorly functioning bowel means sufficient amounts of certain micronutrients can't be

absorbed and nutritional deficiencies or oversensitivity can occur. Anatomically, there is little to be seen, but it is a malfunction of the digestive organs.

- Infectious diseases of the gut, caused by bacteria, like salmonella, by a parasite, like a tapeworm, or by an overgrowth of a yeast, like candida: a microscopic yeast which is naturally present in the gut, but which can cause illness if it is present in abnormal amounts.

- I want to make special mention of Crohn's disease and ulcerative colitis, which cause ulcers and inflammatory reactions in the small and large intestine, respectively. The cause is partly genetic, and autoimmunity also plays a large part.

Gut disorders can be a source of great frustration, for both patient and doctor: there's little that can be done, and there are very few drugs that really help. For this reason, it can make more sense to turn to lifestyle medicine for solutions.

What Does Your Poo Say About You?

The gut is also sometimes called the second brain. It's part of a larger ecosystem, interacting with your body's other systems, such as your endocrine system, fat tissue, your immune system, and, not to forget, your brain. Just think of the link between stress and the need to go to the toilet a lot — the best proof there is that there are neural pathways connecting the gut and the brain.

The gut is an important gatekeeper in your frontline defences. Along with your immune system, it determines who is let through and who isn't, and together they ensure that your whole body stays okay.

In dealing with many physical problems, the gut is often completely ignored, and that's a great shame. It makes no sense to treat obese people for localised centres of inflammation spread throughout their body, without also addressing the state of their gut. After all, a disturbance in the functioning of the gut very often lies at the foundation of the condition in question.

So, it's important first and foremost to examine that microbiome: how strong are the frontline defences in the gut? How many intruders does the gut actually stop? Is perhaps 'leaky-gut syndrome' — or intestinal hyperpermeability, to give it its scientifically correct name — putting added pressure on the immune system?

How can we look into your gut, then? That's simple: by analysing what comes out of it — your stool. Some of the markers that can be tested are:

- Zonulin: the protein that holds gut cells together. It will appear in your stool if the cells of the gut wall are coming apart.
- Calprotectin: a molecule that forms in guts that are inflamed. This molecule is produced in large quantities in people with Crohn's disease or colitis. The presence of calprotectin in a stool sample also indicates that the cells of the gut are inflamed.
- Secretory IgA: IgA is an antibody that's produced by immune cells. Secretory IgA drifts around in the gut and acts like a nightclub bouncer. Excessive amounts of IgA

are an indication that the immune system in the gut has gone into overdrive.

This basic analysis gives your doctor an overall view of how well your gut is functioning as a barrier. More specific follow-up tests can identify which bugs are present in your intestinal system and whether you have a case of bacterial overgrowth. In the small intestine in particular, such an overgrowth of bacteria can cause multiple problems, including bloating, heartburn, flatulence, constipation, or diarrhoea. In medical terms, this condition is called small-intestinal bacterial overgrowth, or SIBO for short, and it provides clear proof that the quality of the bacteria in your gut helps determine how well it functions.

These days, SIBO can be identified with a simple breath test. Since gut bacteria ferment sugars, the test involves administering a certain amount of lactose and glucose to the patient, and then testing the concentration of methane gas and hydrogen in their breath at different times, to give a very precise indication of which bacteria are too prominent. This method can help diagnose bacterial overgrowth, but it doesn't offer a solution to the problem, as it doesn't provide any information about what is causing the overgrowth.

The News in Your Number Two

You can also have a specific microbiome test. This genetic test is not just an everyday stool analysis, and so it's only carried out on special request. But it's a very thorough and useful test that can provide a lot of information.

- It shows how diverse your microbiome is — that is, how many different species it contains. The lower the diversity, the higher the risk to your health and the less well your immune system is functioning.

- The presence of butyric acid is an important indication. When you eat dietary fibre, your good bacteria break down the fibre and ferment it to produce butyric acid. This short-chain fatty acid has two functions. First, it is reabsorbed by the body, returning to the bloodstream and providing energy. Second, butyric acid, or butyrate, has a major influence on the immune system due to its anti-inflammatory effect. Really, this test result is an indication of how much dietary fibre you are eating — or not eating!

- The presence of certain bacteria can be linked to the risk of developing certain diseases. The greater the number of leaks in the wall of the gut, the more the immune system is activated and the greater the chance of developing allergies and hypersensitivities. Autoimmune diseases also have more chance of developing, as do neuroinflammations. I've already pointed out that the brain and the gut are linked by direct neural pathways, so it's only logical that an inflammation in the gut will not stay put, but will spread, possibly throughout the whole body, and even to the brain. When your gut microbiome interferes with the neurotransmitters in your brain, the result can be depression, mood disorders, and hypersensitivity to stress.

- We are also starting to learn more about cancer risks. Caution: I'm *not* saying we can diagnose cancer via the microbiome! But what I *am* saying is that the presence of a certain bacterium in a disturbed microbiome *can* indicate an increased risk of bowel cancer.

- Knowledge of the state of a patient's microbiome can also help make predictions, such as the amount of gas they will produce when they eat certain foodstuffs. An irritable bowel, for instance, will overreact to certain carbohydrates with hyper-fermentation, causing more gas to be produced.

- We can also see links between your response to certain diets and the risk of histamine overproduction, or underproduction of the happiness hormone serotonin.

- Once all these factors have been mapped, one way you can react accordingly in the very short term is by taking supplements. Most important, perhaps, are glutamine to repair your gut wall, probiotics to provide temporary support (e.g. during a course of antibiotics), and butyric acid for its immune-modulating properties. But you can also start to change your long-term behaviour, such as adapting your diet. The body's microbiome takes years to get into such a mess, so you should give it enough time to get back to the way it should be.

Supporting and guiding you in this — is that still the purview of a doctor? Not really. A good nutritionist or health coach can also help you come up with a healthy diet that's tailored to your specific microbiome. And a great piece of general advice is: the more (naturally) colourful the range of food you eat, the better!

Exercise

Phillip is a man of 32, with a young family and a busy career. He wants to do everything to the best of his abilities. Illness doesn't fit in with his life plan, and he came to see me to ask what medication he could take to make sure he stays as healthy as possible. Naturally, that was my cue to trot out my hobbyhorse: the best gift Phillip can give his body is a better lifestyle. Above all, more exercise. Indeed, he tells me that on the days he manages to fit in some sport, he feels great.

Do you know why that is? When you exercise, it's actually like taking a gigantic vitamin pill, chock-full of BDNF. This brain-derived neurotrophic factor is a magic molecule that nourishes your entire nervous system. BDNF stimulates nerve-cell growth, in your brain as well as in your muscles, kidneys, and the entire periphery of your body. Since every part of your body is supplied with nerves, BDNF has an overall beneficial physical effect. Furthermore, BDNF also strengthens your long-term memory.

And exercise is the biggest producer of BDNF!

So, what exactly do I mean by *exercise*? Should we all be completing a ten-kilometre run every day? Should we all be aiming to run the Boston Marathon? No. In general, there's no point in exercising like crazy every Saturday and then lounging on the couch for the rest of the week. That will benefit your health very little.

Leading an active life during the day is the first step in the right direction. Daily activities like vacuum cleaning, walking to the shops, and folding the laundry mean we have a certain amount of calorie consumption for free, so to speak. You can gradually increase that amount towards the '10,000 steps a day' rule, by which time you can really claim to have an active basic lifestyle. Children who exercise a lot during their first few years at school show better overall development, as it makes their brains larger and smarter.

It sounds like a fairy tale: but you will live a longer, happier, and smarter life and will probably get sick less.

But let's get back to Phillip. He's willing to do more exercise, but wants to know what scientific evidence my theory is based on. Well, as a doctor I can clearly show that there's a connection between physical activity and the body's defence system. And that connection can be supported with the right nutrition.

Let's rewind for a moment to 1902, when scientists carried out a study of the runners in that year's Boston Marathon. They discovered that the marathon participants had fewer white blood cells than other people, and further research revealed that the white blood cells of people who regularly do sport behave differently from those of people who aren't regular exercisers. The alert readers among you will now be thinking: *Fewer white blood cells? That must mean their immune system is working less well, right?* You might think that, but what if the cells they do have are simply stronger and more powerful? Let me explain ...

Strenuous exercise causes a temporary deterioration in the functioning of the immune system. This can actually be seen in the blood: the inflammatory markers (such as CRP, which is a protein that indicates inflammation in the blood) rise slightly, temporarily increasing the risk of contracting a respiratory infection, for example. This is the acute effect of exercise: the natural killer cells, B cells, and T cells work less efficiently for a time, and the risk of infection is two to six times greater immediately after. Overly strenuous exercise, such as running a marathon or competing in the Tour de France, severely disrupts the body's defences. Exercising above your training level is also not altogether healthy for your body.

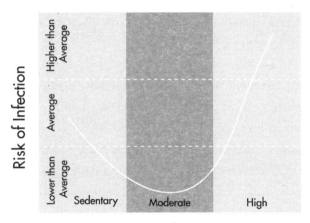

Amount of Exercise

Exercise doesn't mean you have to strain yourself to the point of exhaustion. I'm a man of moderation, and extreme behaviour is never a good idea, irrespective of its effect on the immune system. But if you still want to prepare for a marathon, or a have any other sporting goal in mind, you should make sure you only increase your training by a maximum of 10 per cent per week.

Anyway, be that as it may, moderate physical exercise is ideal if you want to improve your immune system. It starts with basic physical movement: make sure that in general you're reasonably active during the day, and take occasional breaks from sitting down. Feeling a bit like Jesus? Yes? Then: ARISE AND WALK! Actually, do it right now: put this book down and walk twice round the table. Or round the room. Or go up and down the stairs just once. This — minimal — amount of exercise sets your lymphatic fluid in motion, with the result that waste products are removed from your system more quickly. It's better to move a little every hour than to sit around all week on your chair or couch and then go for a three-hour run on Sundays.

If you really want to start exercising, make a training plan that's suited to your needs and follow it regularly. And don't forget to include recovery periods in your plan. If you go for an hour's run, make sure you eat a hearty, healthy meal afterwards, and get enough sleep that night to replenish the energy deficit you created by exercising. And keep a good eye on yourself: if your body sends you messages, such as a sore throat, muscle pain, or shivers, then listen to it!

Exercising with a Fever

If your immune system is busy fighting an infection, you should get plenty of rest to give your defence mechanisms every possible chance to do their job. A simple head cold might not require complete rest, but any infection below your neck should make exercising out of the question. If you're running a fever, it can even be life-threatening: I can't stress that enough. After all, fever, as you may remember, is a sign that your entire army of defence soldiers is in action. The greatest danger is that the infection, and the inflammation circulating throughout your body as a result of it, will attack your heart muscle or your pericardium, the sac of tissue surrounding your heart. The tissue can then break down somewhat — remember the different kinds of cells that shoot and destroy — leaving scar tissue, which can lead to cardiac arrhythmia (an irregular heartbeat) later, as the damaged tissue is unable to work as efficiently as before.

What Is Fever?

Since body temperature fluctuates throughout the day, and is not exactly the same for everybody, it's wise to get a base measurement for your body temperature. This is done by measuring your temperature at different times of the day over the period of a week. The thermometer should be as close to your body's core as possible. That means the most efficient way to measure it is rectally — but measuring it in your inner ear or under your tongue also gets close to the core. Modern, high-quality infrared thermometers are also highly recommended, especially if the thermometer is to be used by more than one person.

The Benefits of Exercise

Now, I've talked mainly about what you definitely should *not* do and what can set your defences back. Why then, for heaven's sake, should anyone do any exercise at all? Phillip is one of the large group of people with quite a healthy diet and lifestyle who don't exercise enough. He's not interested in running marathons, but he does want to find the most efficient way to get enough exercise to keep his immune system working well. And he knows that a training plan including 30 minutes of exercise three to five times a week will result in considerable health benefits for him.

To begin with, exercise has a very beneficial effect on your sleep. When you exercise, you use up a lot of ATP (adenosine triphosphate — the body's energy source at a cellular level) and increase your reserves of adenosine. The more adenosine you accumulate, the greater your sleepiness becomes. And the greater your sleepiness, the better the quality of your sleep!

Exercising activates your sympathetic nervous system, which is your stress system. That's why some people have trouble getting to sleep after exercising late in the evening. But if you exercise regularly, your body will soon learn to compensate for this by stimulating your parasympathetic nervous system, which is your soothing system. Some people see an increase in the feeling of being relaxed after half an hour; others might feel it after two hours. You can find out for yourself when the latest time in the evening is that you can exercise and still go to sleep at the normal time: that's when your two nervous systems are in equilibrium. For the sportier types among us, the soothing system will eventually gain the upper hand, which is why they sleep better and are more resilient to stress.

Although inflammatory markers increase significantly after an exercise session in physically active people and those who exercise moderately, their general level of inflammation is much lower. Active people are 'chronically healthier', which can be proven by certain specific blood tests for the cytokines called interleukin 6, 8, and 10. Interleukins are molecules used by our white blood cells to communicate with each other, as you may remember from the previous chapter. People who exercise have different levels of them, which shows us that exercise lowers the risk of infection over the long term.

The inflammatory markers in the blood of people who are physically fit are considerably lower than those of people who are obese, but they are also far lower than those of people whose weight is 'normal'. The bodies of people with obesity are in a constant state of chronic, low-grade inflammation: their defences are constantly in fight mode. Furthermore, fat cells produce cytokines of their own. The level of activity of physically fit people's immune systems is lower as they don't need to be constantly fighting all kinds of minor conditions. That means their immune system can spring into action much more effectively when there is a real problem. All the little cancer cells our bodies are constantly producing are figuratively and literally nipped in the bud. Obese people's immune systems are kept so busy dealing with all kinds of inflammations that tumour cells can slip through the net and go on to develop further. Cardiovascular diseases are also more common in obese people, as their immune system is less able to clean up their clogged arteries.

When you exercise, you use a lot of oxygen. And when you use a lot of oxygen, you create a lot of free radicals. This is known as oxidative stress, and, when it happens, your immune system is triggered to capture the free radicals. So, exercise initially increases oxidative stress, but, in doing so, exercise also helps develop the mechanisms for dealing with it, resulting in less oxidative stress over the long term. Less oxidative stress, in turn, means less chronic inflammation ... And the good news goes on!

Exercise also has a positive effect on your gut flora and your microbiome, and all the microorganisms that live in your intestinal tract. About a third of your microbiome is the same as everybody else's. The rest is different for everyone, as it depends on all kinds of factors, including diet, lifestyle, age, gender, etc. The metabolites released by physical exercise are extremely nourishing for the desirable kinds of gut bacteria. The more diverse your microbiome, the better your immune system will work. Everything is connected!

Strenuous Exercise

There are additional factors to take into consideration when people train more intensively or cycle, run, or swim for more than an hour several times a week. These people move into the realm of what we call endurance sport. Many of them eat too little when they exercise, which increases their risk of infection.

In the 1990s, scientists discovered that athletes who consume enough sugar while exercising run a far lower risk of extreme stress-hormone fluctuations and are less vulnerable to infection after a sports session. They got sick far less frequently when they ate up to 60 grams of carbohydrates per hour of exercise. That's about the maximum amount a normal person can digest in an hour, yet it's a mere fraction of the energy an athlete expends while training: each gram of carbohydrate provides four calories of energy, and running or cycling can easily consume between 600 and 1,000 calories an hour. Even with the maximum intake of 240 calories an hour, it will never be possible to compensate for that energy consumption through food.

If you do very strenuous exercise, you need to replenish your energy reserves with carbohydrates, otherwise you will push your body too far towards the brink, and that will in turn increase both your stress-hormone levels and your risk of infection.

The carbohydrates I'm talking about don't necessarily have to be the rapid sugars that competitive cyclists use during long-distance races. Bananas and dried fruit are also rich in carbohydrates, and they contain metabolites that have a positive effect on levels of hormones such as serotonin and dopamine. The polyphenols they contain play an important part in inflammatory responses after exercise, and they also provide a lot of antioxidants, which accelerate post-exercise recovery even further. So, it's very important to eat and drink when exercising for a very long time. I know several marathon runners who even stash food or drink at various locations along their training route.

Furthermore, there is a demonstrable connection between chronic inflammation, physical exercise, and immunosenescence.

Immunosenesc— What??

Immunosenescence is the ageing process of the immune system. The older we get, the older and less resilient our immune system becomes. Our first, second, and third lines of defence begin to function less well, which automatically puts older people in a higher risk group. Not just for COVID-19, but for any potential infectious disease. Immune cells age like any other cells in our body. They become weaker and less lively. The older our cells get, the more at risk we are from infections, autoimmune diseases, and also cancers, which develop because our defences become less active in detecting the mini-tumours they have to clear up every day.

The more active older people remain, the fitter their defence cells will stay, the higher the quality of their T lymphocytes will be, and the less quickly they will become sick. Exercising regularly is a very sensible strategy for older people. Of course, you don't have to be able to complete a mountain stage of the Tour de France: moderate physical exercise and an active lifestyle are more than enough. More physical exercise also brings benefits when it comes to vaccinations: older people who are physically active have a better immune reaction to the flu vaccine, which means they are better protected by it.

To help you understand our cells' ageing process better, I need to tell you a little about telomeres. They're the ends of the chromosomes, which are present in every cell of your body — including your white blood cells. Your cells are constantly dividing throughout your life, in order to survive. Yet every division shortens the telomeres slightly, until they eventually disappear altogether. At that point, the cell is dead: it can no longer divide to survive. An active lifestyle slows down that telomere-shortening process, extending the time they can keep functioning before they get worn out.

The Four Components of Exercise

There are four components to good exercise: stability, strength, stamina, and high-intensity training. You need to pay enough attention to each of these four elements if you want to exercise in a healthy way.

Without stability and strength, it is impossible to exercise safely. Your risk of injury is ten times higher if you lack core stability. As babies we are born with a pretty good foundation — just watch how small children move: when they crouch down, they execute a perfect squat, naturally flexing their knees in the safest way. Children naturally get from A to B by running rather than walking — that's their instinctive way of moving. The older we get, the more we lose those positive qualities, because we move too little and sit too much. This is why exercises that increase core stability should be a standard element of any training plan. Because acquiring core stability is so important, it makes sense to consult a good physiotherapist or personal trainer. Or just observe any two-year-old who's just learned to walk and still instinctively moves in the naturally correct way.

The second important component of exercise is strength. Stronger muscles move more mass faster. Strength training isn't just for bodybuilders, and it doesn't necessarily mean you have to go and pump iron in a gym. There are a huge number of excellent apps for basic strength training, with short workouts lasting just seven minutes, for example. And the only equipment you need is a mat and a wall! A slow build-up is also important here, to avoid direct strain injuries and to help keep you motivated. And when it comes to anti-ageing effects, strength training is nothing but beneficial: the more active you want to remain, the more muscle mass you will need. Furthermore, strength training makes your bones stronger, reducing your risk of osteoporosis and breaks.

Exercise for stamina and high-intensity training are referred to in the specialist literature as zone 2 and zone 5 training, respectively. That probably doesn't mean much to you, right? *Zone 2* and *zone 5* refer to your lactate production. When your muscles move, you burn sugar or glucose. The sugar is turned into energy, as I described earlier, and your cells start producing ATP. There are two routes your body can go down to produce ATP: one with oxygen, the other without.

If you go out for a leisurely hour-long bike ride, you can keep producing large amounts of ATP over an extended period while you're using up oxygen, which means you can carry on exercising for a long time. You are now in zone 2, or the aerobic zone. If you suddenly ride up a steep hill and have to really push to get to the top, your body switches to zone 5 and begins converting sugar into ATP without using oxygen — that is, anaerobically. But you can only do that for a limited period. Lactic acid, or lactate, forms in your muscles, the level of acidity increases, and you begin to feel the burn.

Nowadays, you can train both systems at the same time, so that the body's recovery mechanisms are put to work faster. This type of exercise is called high-intensity interval training, or HIIT.

It's important to train both systems properly. Zone 2 is important for keeping your cardiovascular system fit. Your heart rate increases a little, but not too much, which strengthens your cardiac muscle. Blood is pumped round your body more quickly, which improves the elasticity of your blood vessels. And now we're getting to the point: when your entire circulatory system is stimulated, your lymphatic system also functions better. The lymphatic system, if you remember, is the body's sewage system. All white blood cells that are fighting an infection send toxins to your lymphatic vessels to be disposed of. That doesn't happen automatically! Our lymphatic system isn't like our cardiovascular system, in which blood is pumped round the body by our heart, round and round and round. The toxins sent to the lymphatic system follow a one-way street to our lymph nodes, but there's no pump to get them there; it's the contraction of your muscles that helps the lymphatic system defy gravity. Once at a lymph node, the toxins are removed and the clean fluid can return to the body via the bloodstream.

By the way, you can easily help your lymphatic drainage along by treating yourself to a foot massage, or by joining in with your kids' jumping games, like skipping or trampolining, or by visiting an osteopath, who can release certain lymphatic blockages. Keeping hydrated also helps maintain better circulation!

Sleep

Celine is 38, is the mother of two small children, and works as an intensive-care nurse. Her husband is self-employed, very career-oriented, and rarely at home. It's mainly Celine who takes care of the children, with the help of her mother whenever she has to work the late shift. She has no significant health problems, but she is a little overweight. During her night shifts in particular, she tends to eat unhealthily. She'd like to start intermittent fasting, both to improve her health and to lose a few of those extra kilos, but she doesn't know how to get started, due to her irregular lifestyle. She'd also like to do more exercise, but she doesn't have the time. She sleeps very badly as her biorhythms are so irregular. She puts herself under a lot of pressure because she wants to be good at her job, as well as being a perfect mother and partner.

Her husband, Thomas, is 46 and well on his way to building a rewarding career with his own business. But it's quite stressful and he often lies awake at night. He sometimes has panic attacks at 3.00 am. He's tried meditation to manage his stress, but he soon realised it's easier just to take a sleeping pill or down a couple of glasses of wine. During the day, he has to give so much of himself that by the evening his batteries are empty and all he can do is flop into an armchair. Unless, that is, he has to take clients out to dinner, which is a regular occurrence. At the weekend, he often doesn't have the energy to play with his children. He tries not to fall asleep in his armchair, but that usually backfires and he stays up far too late. He then has trouble getting up in the morning, and drinks too much coffee. He often tells himself, 'Tonight, I'll get an early night.'

We all suffer from a chronic lack of sufficient sleep. During the — almost — global lockdown, one sleep-tracking company monitored the sleeping habits of 68,000 people. That study showed people in several counties slept up to 20 per cent more. When we're under less work and social pressure, it seems we listen more closely to our biological clock and sleep more.

People also experienced far less social jetlag, which is the difference in the amount of sleep we get at the weekend compared to during the week. In pre-COVID times, we all slept little during the week, because we had to get up early to go to work and had social obligations keeping us up late at night; then we lay in longer at the weekend, to try to catch up on missed sleep. During the pandemic, as we stayed home en masse, we all switched to permanent weekend mode.

Under normal circumstances, we often live with an abnormal biorhythm. External pressure means we get less sleep and accumulate a massive sleep deficit. When we are able to sleep more during the week, we have less need to catch up over the weekend. But anyway, science tells us that catching up on sleep is an illusion.

On average, we all went to bed 30 minutes later than usual during lockdown, and got up 50 minutes later. There was a real shift in our sleep duration when we were more able to follow our natural biorhythm.

Despite this, a large proportion of those 68,000 people claimed that they were sleeping less well than normal. However, to a large extent, that was linked to work-related worries and the fear of getting sick.

Morning Person or Night Owl?

Just as what you eat and when you eat it are important, it also matters *when* you sleep. To use a modern buzzword: during lockdown, we paid more attention to our inner chronotype.

Scientist love to pigeonhole everything, so it's not surprising they've come up with four sleep types to categorise people into. There are extreme morning people, normal morning people, mid-evening types, and true night owls. Which chronotype we each fit into is partly genetic, as we all have our own internal clock, which, among other things, regulates our circadian rhythm (sleeping–waking cycle). Morning people jump straight out of bed and are fully alert in the morning. Their productivity falls continuously as evening approaches. Evening people, on the other hand, would prefer to remain buried as deeply as possible under the bedcovers in the morning, but feel a peak in their energy levels in the evening.

Your chronotype evolves as you get older: babies sleep round the clock, only waking up every now and then. Young children, on the whole, are real morning people, while teenagers go to bed later and get up later. When we reach adulthood, our bedtime gets a little earlier again, and we tend to get up a little earlier, too. Unfortunately, we have to comply with social demands and norms, which sometimes prevent us from following our biology's lead. Secondary schools, for example, should really switch to teaching hours of roughly 9.30 am to 6.00 pm. That would be far better adapted to the sleep rhythms of adolescent students.

Celine's sleep rhythm is also disturbed by her night shifts, which means she no longer really knows whether she's a morning person or an evening person.

Below is the standard morningness–eveningness questionnaire, or MEQ, which determines your chronotype based on your answers to a list of questions.

How It Works
Read each question carefully. Answer as honestly as you can. Don't go back to questions you've already answered — your first response is usually the most accurate one. Don't skip any questions.

The Questions

- What time would you get up if you were entirely free to plan your day?

5.00 am – 6.30 am	5
6.30 am – 7.45 am	4
7.45 am – 9.45 am	3
9.45 am – 11.00 am	2
11.00 am – noon	1
Noon – 5.00 am	0

- What time would you go to bed if you were entirely free to plan your evening?

8.00 pm – 9.00 pm	5
9.00 pm – 10.15 pm	4
10.15 pm – 12.30 am	3
12.30 am – 1.45 am	2
1.45 am – 3.00 am	1
3.00 am – 8.00 pm	0

- If you have to get up at a specific time in the morning, how much do you depend on an alarm clock?

Not at all	4
Slightly	3
Somewhat	2
Very much	1

- How easy do you find it to get up in the morning (when

you are not awakened unexpectedly)?

Very difficult	1
Somewhat difficult	2
Fairly easy	3
Very easy	4

- How alert do you feel during the first half-hour after you wake up in the morning?

Not at all alert	1
Slightly alert	2
Fairly alert	3
Very alert	4

- How hungry do you feel during the first half-hour after you wake up?

Not at all hungry	1
Slightly hungry	2
Fairly hungry	3
Very hungry	4

- During the first half-hour after you wake up in the morning, how tired do you feel?

Very tired	1
Fairly tired	2
Fairly refreshed	3
Very refreshed	4

- If you had no commitments the next day, what time would you go to bed compared to your usual bedtime?

A little or no later	4
Less than one hour later	3
One to two hours later	2
More than two hours later	1

- You have decided to do physical exercise. A friend suggests that you do this for one hour twice a week, and the best time for him is between 7.00 and 8.00 am. Bearing in mind nothing but your own internal clock, how do you think you would perform?

I would be in good form	4
I would be in reasonable form	3
I would find it difficult	2
I would find it very difficult	1

- At what time in the evening do you feel tired and, as a result, in need of sleep?

8.00 pm – 9.00 pm	5
9.00 pm – 10.15 pm	4
10.15 pm – 12.45 am	3
12.45 am – 2.00 am	2
2.00 am – 3.00 am	1

- You want to be at your peak performance for a test that you know is going to be mentally exhausting and will last two hours. You are entirely free to plan your day. Considering only your internal clock, which one of the four testing times would you choose?

8.00 am – 10.00 am	4
11.00 am – 1.00 pm	3
3.00 pm – 5.00 pm	2
7.00 pm – 9.00 pm	1

- If you got into bed at 11.00 pm, how tired would you be?

Not at all tired	1
A little tired	2
Fairly tired	3
Very tired	4

- For some reason you have gone to bed several hours later than usual, but there is no need to get up at any particular time the next morning. Which one of the following are you most likely to do?

Wake up at usual time, but not fall back asleep	4
Wake up at usual time and doze	3
Wake up at usual time, but fall asleep again	2
Not wake up until later than usual	1

- One night you have to remain awake between 4.00 am and 6.00 am to carry out a night watch. You have no time commitments the next day. Which one of the alternatives would suit you best?

I would not go to bed until the watch is over	1
I would take a nap before and sleep after	2
I would take a good sleep before and nap after	3
I would sleep only before the watch	4

■ You have to do two hours of hard physical work. You are entirely free to plan your day. Considering only your internal clock, which one of the following times would you choose?

8.00 am – 10.00 am	4
11.00 am – 1.00 pm	3
3.00 pm – 5.00 pm	2
7.00 pm – 9.00 pm	1

■ You have decided to do physical exercise. A friend suggests that you do this for one hour twice a week. The best time for her is between 10.00 and 11.00 pm. Bearing in mind only your own internal clock, how well do think you would perform?

I would be in good form	4
I would be in reasonable form	3
I would find it difficult	2
I would find it very difficult	1

■ Suppose you can choose your own work hours. Assume that you work a five-hour day (including breaks), your job is interesting, and you are paid based on your performance. At what time would you choose to begin?

4.00 am – 8.00 am	5
8.00 am – 9.00 am	4
9.00 am – 2.00 pm	3
2.00 pm – 5.00 pm	2
5.00 pm – 4.00 am	1

- At what time of day do you usually feel your best?

5.00 am – 8.00 am	5
8.00 am – 10.00 am	4
10.00 am – 5.00 pm	3
5.00 pm – 10.00 pm	2
10.00 pm – 5.00 am	1

- One hears about morning types and evening types. Which one of these types do you consider yourself to be?

Definitely a morning type	6
Rather more a morning type than an evening type	4
Rather more an evening type than a morning type	2
Definitely an evening type	0

Score

Add up all your points and write the total in the box below:

Your score can vary between 16 and 86. If you scored 41 points or fewer, you are an evening type. A score of 59 or more indicates that you are more of a morning type. If you scored between 42 and 58 points, your chronotype is neutral.

When You Fight Biology, You Normally Lose

Why is it important to know your chronotype? Sometimes, life gets in the way of an ideal sleep pattern, as is the case for Celine. Often, we are so busy that we don't even know what time of day we function best at. Finding out your chronotype can help you organise your life better. If you're an evening person, you can arrange things as far as possible so that you can start your day later and don't have to get up so early, and you can work for longer in the evening if that's when you perform best. Evening people who are forced to start their day at 6.00 am often suffer from long-term sleep disorders.

The sleep study during lockdown also revealed something else: people were dreaming more! Let's drift away together into the land of dreams.

There are two kinds of sleep: REM sleep, which is extremely light and is the phase in which dreaming takes place, and non-REM sleep, which passes through four phases from light to very deep. During sleep, your brain alternates several times between REM and non-REM phases. Each cycle of REM to non-REM sleep lasts about 90 minutes. In the first half of the night, your body predominantly focuses on non-REM sleep — it goes into a state of deep rest. Your body needs the second half of the night for recovery and processing, so more time is spent in REM sleep. That's why you tend to dream more towards morning. It's a great combination, actually! But if your sleep pattern is regularly disturbed, by worry, as in Thomas's case, or by night work, as in Celine's, the whole system can get out of kilter.

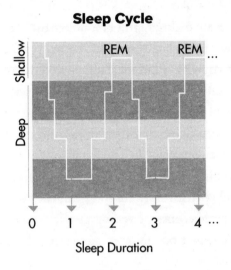

Sleep Cycle

Sleep Duration

What happened during the pandemic? In general, we slept 50 minutes longer in the morning: that means we didn't extend the rest phase, but got more REM sleep cycles in, and so we dreamed more.

Another reason why the coronavirus crisis increased the amount that people dreamed is the need to process negative emotions and stress. Dreams are extremely important for emotional health: they are a form of nocturnal therapy. We increase our mental resilience by dreaming. The energy field in our brain, which has been stimulated all day, gets the time to rest, and our brainwaves become less active. It's often a good idea to sleep on an important decision, as it gives your brain a chance to clear out all the waste products, including mental ones. You can then make new associations, which gel in your brain through the night. REM sleep is emotional first aid!

Back to Thomas and Celine. In the evening, when Celine is working the late shift, Thomas lies on the couch, exhausted. He's had to make decisions all day and now has no energy left to do anything more than watch TV. But this exposes him to far too much blue light, which his brain registers as daylight. His body then doesn't produce enough of the sleep hormone melatonin and he can't get to sleep. If he drinks a glass of wine, he has the feeling of having slept well and being refreshed in the morning. One glass of wine soon becomes two, and then, without even really noticing it, he's soon polishing off an entire bottle every evening.

What happens when you drink alcohol? Your breathing rate increases, because your body wants to get rid of the alcohol, your heart rate goes up, and your body temperature rises. At the same time, your heart-rate variability goes down. Heart-rate variability is the difference in time between two heartbeats, and high variability is a good thing. The less equal the time between two heartbeats is, the calmer we are — so there's an important correlation between heart-rate variability and stress. If you feel perfectly balanced, the variability between your heartbeats will be very high.

Our friend Thomas has the feeling that he sleeps well after downing a bottle of wine, but he has simply knocked himself out and interfered with his natural sleep architecture. His body and mind don't recover during the night, as he has both hindered his REM sleep and activated his stress system. His brain gives him less emotional support, so he wakes up feeling unrefreshed and spends the day under stress, only to begin the whole cycle again in the evening.

After consulting his doctor, Thomas decides to give up alcohol and starts wearing glasses with a blue-light filter in the evenings. And since then, he has been sleeping a lot better! That positive reversal has also had an important effect on his immune system. We know that sleep deficit can have serious consequences, and studies have shown that people who regularly get less than seven hours' sleep have more chance of catching a common cold, and people who sleep for less than five hours a night are more likely to contract pneumonia.

The amount of sleep we get even influences our antibody response. One study shows that people who sleep too little in the week before they receive a vaccination produce fewer antibodies in response to the shot.

Celine no longer has a normal biorhythm due to her irregular work schedule, and she has also completely disrupted her body's production of two important hormones: melatonin and cortisol.

Circadian Rhythm
Daily Cortisol and Melatonin Cycles

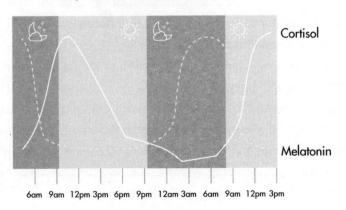

6am 9am 12pm 3pm 6pm 9pm 12am 3am 6am 9am 12pm 3pm

Melatonin, the sleep hormone, and cortisol, the stress hormone, are produced in rhythms throughout the day. The later the evening gets, the more melatonin circulates in the bloodstream, peaking at the point when you can no longer stay awake. It falls again towards morning, and your melatonin production is almost nil during the day. Cortisol does the opposite, peaking when you get up and falling off to almost nothing by the time you should be going to bed. Melatonin sends you to sleep, cortisol makes you alert. A good balance between these two hormones is essential for good-quality sleep. And there is a third substance involved here: adenosine, which forms the basis of the energy molecules in your body. The longer you stay awake, the more energy you will have used, and therefore the more adenosine will be released, which is your body's signal to you that it's time to turn in for the night. This is what makes you feel increasingly sleepy in the evening.

For Celine, it's important that she pay attention to her production of these three substances. An afternoon nap that somewhat lessens her feeling of sleepiness will have a positive effect on her cardiovascular system and on her learning and memory capacity. By contrast, sleeping during the day is precisely the thing that should be avoided by Thomas, who often has trouble sleeping at night, as it may mean that he's not tired enough to go to sleep by bedtime. For Celine's production of melatonin, which in turn stimulates the production of the happiness hormone serotonin, it's vital that she eats enough food that's rich in tryptophan. Foodstuffs that contain this amino acid include chocolate, oats, dried dates, milk and yoghurt, red meat, eggs, chicken, and chickpeas.

None of this works like a sleeping pill, but it does provide the body with everything it needs to produce the optimum quantities of melatonin and serotonin.

As a shiftworker, Celine doesn't always have the opportunity to eat healthily, but she can support her body's production of melatonin somewhat by taking dietary supplements. If you get home at six o'clock in the morning and have to start your 'night' just as your production of melatonin is almost zero, you might well need some help with getting to sleep more quickly and with getting back to your normal rhythm afterwards.

To a certain extent, taking magnesium can also have a calming effect on your body when you're struggling with sleep disorders. A warm bath with Epsom salts, or magnesium sulphate, can work wonders. Of course, my motto is 'measure to manage', so you may want to ask your doctor to test your blood for a magnesium deficiency.

Sleep-Restriction Therapy

Those who have a real sleep disorder will not be able to remedy it by eating a bit more healthily or popping a melatonin pill. Doctors will often prescribe 'real' sleeping pills, but that can't be a long-term solution, especially in the light of lifestyle medicine.

But sleep-restriction therapy can provide a solution. It's based on the principles of behavioural therapy. It involves building up as large a sleep deficit as possible by restricting the duration of a patient's sleep. Instead of crawling into bed at 11.00 pm, the patient stays awake until two o'clock in the morning, but still gets up at the normal time of 7.00 am.

That window for sleeping is barely five hours, which is actually too little, but, since the patient has accumulated such a large sleep deficit, their quality of sleep will be that much higher. The hours the patient actually spends in bed are hours of efficient sleep, rather than hours spent tossing and turning.

People in some pre-industrial societies, who are isolated from the outside world and live principally from hunting, have a natural tendency to be inactive in the afternoon, as that's when their alertness is at its lowest. We, too, show evidence of a natural programming that makes us less focused in the afternoon.

In comparison with humans, monkeys have far less REM sleep. REM sleep reduces muscle tone — and when a monkey has no muscle tone, it falls out of its tree.

Routine Isn't Always Boring

We take it for granted that young children need a regular bedtime ritual to help them understand when it's time for them to sleep and to trigger their feelings of sleepiness. Once their bedtime routine is set in motion, they immediately start yawning and realise it's time to settle down for the night. As adults, we lose that routine, but we would all benefit from creating a bedtime ritual for ourselves to ensure an optimum sleep architecture. Let me give you a few tips to help you do that.

- Go to bed at the same time every night, and get up at the same time every morning. Regularity is your best friend.
- Create a nice ritual for yourself. For example, put your pyjamas on an hour or so before your bedtime and drink a cup of tea.
- Avoid blue light by reading a book rather than watching television, or listen to some relaxing music.
- If you do watch TV, don't opt for an exciting series that stimulates your stress system.
- Avoid all sources of light in your bedroom, even from an alarm clock. This will also help you avoid anticipatory anxiety, which often occurs when you wake up in the middle of the night, see the time, and worry about how few hours you have left to sleep (because it will inevitably never be as much as you'd like).
- Put. Your. Phone. Away.

Stress

Single mum Natalie is 41, has three children, and has been divorced for four years. At work, she was recently promoted to a position with more responsibility, in preference to her colleague Rob, who also wanted the job. He has taken it badly and regularly tries to make Natalie's life difficult. He often transfers work to her that's not actually part of her job. Natalie is a people pleaser and just wants to do right by everybody. Besides, she's determined not to let this guy get the better of her! When she gets home from work in the evening, she has three children demanding her attention. She loves spending time with them, and they are her entire motivation, but they consume the last bit of energy she has left ... At some point, she feels she's constantly running on empty and is losing control of the situation. She can slam on the brakes, but how can she handle that amount of stress?

The Elephant and the Mouse

The difficult thing about stress is that it can't really be measured and can't even easily be defined. What's highly stressful for one person might be a breeze for another. Various acute stress reactions can trigger a state of chronic stress. There is no objective yardstick. All we can really do is to chart the consequences of stress by measuring the amount of stress hormone produced by a person, how much adrenaline they have in their blood, how fast their heart is beating, what their heart-rate variability is, and, most importantly ... how they perceive it themselves.

Fortunately, modern medicine has finally begun exploring the link between body and mind, and now many scientists are busy researching the connection between stress and the immune system.

Negative stress occurs when there is an imbalance between your stress capacity and the stress load. Imagine an elephant and a mouse on a seesaw: if you are the elephant and your stress is the mouse, the balance is in your favour. Conversely, if you're the mouse and the elephant at the other end of the seesaw is your stress level, you have a big problem. That's when it's time to start seeing the aspects of your life as fitting into two categories: energy guzzlers and energy providers.

Natalie sits down to have a stern word with herself about regaining balance in her life. Her children, those little energy guzzlers, are also energy providers, since they are after all what motivates her to keep going. Her work gives her energy, but the difficult situation with her colleague means it also consumes a lot of her energy. Eventually, she decides that she will keep every Thursday evening free for a sauna visit with her best friend, no matter how busy she is at work or how little time she has been able to spend with the kids.

The Anatomy of Your Brain

Natalie presses the pause button every Thursday evening, and that has a great positive effect on her immune system. Let me take you now on a trip inside the head of Professor Brain, to show you just how that organ works.

The brain is the body's electrical system. Each cable in the network is a nerve cell, and each nerve cell is connected to another via a synapse, which is like a connector clip joining two wires in your home.

A synapse is actually a gap across which certain substances, or neurotransmitters, can carry signals from one nerve cell to the other. The most important neurotransmitters are dopamine (the reward hormone), serotonin (the happiness hormone), adrenaline (the fight-or-flight hormone), glutamate (the general-purpose neurotransmitter), and GABA (the calming neurotransmitter). The wiring in the brain is dynamic: certain synaptic connections can be rerouted. This concept is known as neuroplasticity. People who relearn to do everything, from walking to talking, after suffering brain damage in a serious traffic accident are an excellent illustration of the power of neuroplasticity.

It's All in the Mind

You can also rewire your brain to become a better version of yourself. Let me explain:

Natalie has a very busy Wednesday, but she has planned the day carefully, since she needs to rush from one activity to the next and doesn't want to get behind schedule. Everything goes according to plan until she suddenly remembers that she still has to buy a present for her youngest daughter to take to the birthday party she's been invited to ... In Natalie's head, the cogs start turning in overdrive: the nerve cells spring into action and start firing off signals — the brainwaves. Natalie's body connects these signals with a physical reaction. Her endocrine system thinks, 'Uh-oh, there's a huge mountain to climb'; her body reacts with a shot of adrenaline and a little extra stress hormone; her heart rate increases. Natalie experiences a physical awareness of the stress she's under.

This reaction has its origin in the time when we were still living in the jungle and might suddenly encounter a sabre-toothed tiger — the adrenaline rush that this encounter would trigger helped us to run away from the tiger as fast as humanly possible.

Another important element of the reaction generated when we fire off brainwaves is emotion. Every thought elicits a biological response, which is followed by an emotion. That emotion can be happiness, or, when we are under too much stress, it can be anger. Our thoughts become our behaviour, and then our emotions also become our behaviour.

If you can gain control over your thoughts, you can also control your emotions. It's about rewiring your brain, remember?

Okay, Doctor, let's do it! How do I reroute those wires? I could certainly do with having a bit less stress in my life ...

Well, it's not quite that simple, I'm afraid. To really become a better version of yourself, you have to make adjustments at the level of your thoughts. To regulate your emotions, you have to control your memory. Or at least become aware of your thoughts.

Very well, Doctor, that doesn't sound too difficult. I'll just meditate a bit, shall I?

Don't Let Your Thoughts Control You

Meditation is indeed a panacea. The difficulty lies in our thoughts. Only 5 per cent of our mental activity is conscious, the other 95 per cent happens at the unconscious level.

Let's return to Natalie's story, at the point in her super-busy Wednesday when she realises she still has to buy a birthday present for her daughter's little friend. She could solve the problem by quickly picking up a gift voucher from her local toyshop, but neither she nor her daughter would really be satisfied with that solution. She feels guilty and doesn't want to upset her daughter, so she quickly rings round some of the other mums, and they eventually decide to club together to buy a joint present. What started as forgetting to buy a birthday present and a potentially disastrous afternoon resulted in a much better gift for the birthday girl, and a sense of pride in a lot of the other party guests because they helped make their friend so happy.

And Natalie? She's still stuck in her unconscious emotions: her 5 per cent rational thoughts tell her she dealt with the problem successfully, but they are drowned out by the other 95 per cent, which are still raging madly, giving her a bad feeling for the rest of the day.

Just as you can't boost your immune system by simply pushing a button, you can't switch your emotional and hormonal responses on and off at will. But meditation *can* help break the cycle of thoughts and feelings that are at the root of certain behaviour patterns. If you break those patterns often enough, you will feel the 5 per cent grow, you will be less prone to becoming a victim of your unconscious thoughts, and you will be able to control your own mood, mindset, and happiness more quickly.

The Monk and the MRI

But, Doctor, are you really telling me I should start meditating?

Yes, indeed, dear reader. I'm a great believer in meditation myself, and I can even prove scientifically that mediation is a medical good-news story. MRI monitoring of the brain activity of a monk while meditating revealed, at the microscopic level, which areas of his brain were active while he was practising meditation. This technology allows us to measure very precisely which brainwaves are active and which are inactive at any given moment.

There are four frequencies of brainwaves. Some pulsate very quickly, while others are super-slow. Scientists call the four types alpha, beta, delta, and theta waves. At birth, 100 per cent of our thoughts are unconscious. If you've ever wondered what a newborn baby is thinking, the answer is: nothing.

Between birth and the age of two, the human brain mainly produces delta waves, which are relatively low-frequency electromagnetic waves.

So, what happens when children turn two? Their brainwave frequencies speed up, becoming theta, or hyper-focus, waves. Small children live in an inner, abstract world of fantasy, without even a nuance of critical thinking. They are not fully capable of thinking rationally until about the age of six. You can tell them anything, so to speak, which is why they will readily believe that Santa Claus exists and can fly through the night sky in a reindeer-drawn sleigh. This is also the phase in which our children are most susceptible to learning many positive things from us!

From the age of six, children's brainwave patterns begin to change again, switching now to alpha waves. Their analytical mind starts to develop, and this is when children gain the ability to analyse and interpret things independently and to draw their own conclusions for the first time. They begin to combine the external, analytical world with their own fantasy world. If you ask them to be a bear, they don't *pretend* to be a bear; in effect, they *are* a bear. And if you attach fairy wings to their back, they aren't *playing* a fairy, they *are* a fairy.

Between the ages of eight and 12, children's brainwave patterns change once again, this time to beta waves. These are the most common brainwaves in adults. They indicate an awake, alert state. The more stressfully you interpret things, the more deeply entrenched in beta brainwaves you are.

Most common, you say, Doctor?

Indeed. When you're sleeping deeply, you produce delta waves. Between deep sleep (delta) and high alertness (beta), there are alpha and theta waves. We are in alpha-wave territory when we enter a meditative state. The easiest time to achieve this is in the early morning. Every morning, just before you wake up, you experience a kind of rebirth: you are in a very restful alpha-wave state, in which your senses are still unable to perceive very much. Then come the theta waves — a kind of twilight state in which you're not completely asleep and not yet fully awake.

If you can seize that moment, you'll find your thoughts will become super-clear. It's the ideal moment for quiet meditation. Leave your telephone be; don't jump out of bed to start checking your emails or reading the paper.

Don't Hurry & Evacuate, Sit & Meditate

There are many different approaches to meditation. You can practise breathing and 'letting go' with mindfulness exercises. The first step is to alter your brainwave pattern from the beta state to a lower frequency. Without this step, there's no point in trying to silence your thoughts — it will never work! But once you've been able to put yourself into a lower-frequency state, you can just let your thoughts go. Accept, acknowledge, and let go. Choose a position to meditate in — it doesn't matter whether that's sitting, lying down, or hanging from the branch of a tree — as long as you're comfortable. Choose an environment that relaxes you and that makes you feel you're in a position to turn off your beta waves and switch over to alpha waves. Some people use a playlist of music with the same frequency as alpha brainwaves — certain vibrations that aren't too prominent.

This step, switching from beta to alpha waves, is not easy. For me, the best time to meditate is early in the morning, or just before bedtime. Focus on your breathing, focus on the air flowing into and out of your body. Learning to let go takes time. Just as a weightlifter won't be able to lift a hundred kilos on day one, meditation and mindfulness also need to be trained every day. Only then will you reach a point where you realise you can consciously control your thoughts during a hyper-stressful situation which would normally send your brainwaves flying off into beta territory. Learn to listen to how your body feels during stressful situations, so you can recognise that feeling again in the future. 'I have a pain below my ribs' = 'Oh, yes, I'm feeling stressed again'. When you learn to recognise this behaviour pattern, you will realise much more quickly that your unconscious is tricking you with your thoughts, emotions, and behaviours.

Like Natalie, draw up a personal vision for yourself. What kind of parent do you want to be? Do you want to be a reliable employee, a good friend? Who do you want to be? Start from there to create an ideal image of the best possible version of yourself. Once you've learned how to return your brain to the alpha-wave state, you can gently begin to analyse whether the person you are at that moment, and the feelings you are experiencing, fit in with that ideal image of yourself. Channel your own emotions; don't burden anyone else with them. You have to have faith in yourself, to learn that your life is a process of growth and to just accept it. Try to be the best possible version of yourself.

And with these beautiful, ethereal words I close this chapter on lifestyle. Now it's time to take a look at what's inside the medicine cabinet.

'Let's pay the farmer,
not the pharmacist'

Paul Coletta

INSIDE THE

MEDICINE

CABINET

4

You now know — almost — everything about the range of micro-creatures that exist, how your defence system works, and how your lifestyle can affect it. At every stage, your immune response also receives internal support from certain micronutrients.

These micronutrients are an essential part of the immune system. Vitamins, minerals, and phytonutrients ensure that the white blood cells and the cells of the respiratory tract, mucous membranes, and gut are, and remain, healthy and of high quality, so they can continue to do the best job possible in fighting off pathogenic germs. Micronutrients help these cells to function, and keep dividing and developing into specific types of cells, such as phagocytic cells or antibody-producing cells.

Which micronutrients do we need? It's been scientifically proven that vitamins A, D, C, E, and B_6, as well as zinc, iron, copper, and selenium play an important part in the proper production of cytokines, the communication substances needed in various phases of our immune response, and to offer extra support for the inflammatory response as a whole.

What's more, your immune cells are in the direct firing line and require extra protection because of it. They need an efficient antioxidant system to capture the harmful free radicals released by immune processes, and the necessary elements to process and excrete the waste. Vitamins C and

E, zinc, iron, magnesium, copper, and selenium support that antioxidant process.

The Ideal Amounts

To function well, your body needs to maintain the optimum levels of micronutrients. The reference values are calculated by health authorities, with guidelines on the recommended daily intake to avoid deficiencies. I delved a little deeper into biochemistry and physiology as part of my basic medical training. That, in combination with my practical experience in the world of sports medicine, and supplemented by sources from the field of functional and orthomolecular medicine, led me to discover that those recommendations are not always sufficient for optimum health.

Vitamin/Mineral	RDI	Amount Recommended in Complementary Sources
Vitamin A	2,000–2,500 IU	2,000–5,000 IU
Vitamin B$_{12}$	4 mcg	10–2,000 mcg
Folic Acid	200–300 mcg	400 mcg–4 mg
Vitamin C	110 mg	500–3,000 mg
Vitamin D	400–600 IU	2,000–10,000 IU
Chromium	no recommendation	75–200 mcg
Selenium	70 mcg	100–200 mcg
Zinc	8–11 mg	15–50 mg

RDI (recommended daily intake; based on the Belgian High Council for Health, 2015) compared to the recommended amounts from other, complementary sources (orthomolecular medicine, functional medicine). Note the large discrepancies.

Let's take a closer look at vitamin D, as an example: according to the Belgian High Council for Health, the optimum daily intake is 400–600 IU. An increasing number of scientific sources now recommend we aim for a blood value of 50 or 60 mmol/L. This target is difficult to achieve with the recommended daily intake, but it can be attained with a daily intake of 2,000 IU or more. That's around four times as much! Who's right? The truth, as it so often does, lies somewhere in the middle. And this brings us back to my old adage: measure to manage.

Dosages will differ depending on what your aim is. Are you taking supplements for preventive or therapeutic reasons? To repair your immune system or to optimise it? Adapting reference intake values to your personal circumstances is the key to success! And that's the wonderful thing about medicine: we can combine scientific knowledge and guidelines with personal clinical experience and the needs of the actual person sitting in front of us. After all, the phenomenon of biochemical individuality is real. You might require amount A, while I need to take in amount B to achieve the same result. It can depend on genetic factors, as with haemochromatosis — a condition in which iron builds up in the body and slowly poisons the patient's liver. Those who suffer from this should certainly avoid taking iron supplements!

We should no longer be regarding micronutrient therapy as 'alternative medicine'. We should be asking how this form of treatment can be integrated into existing medical procedure. Micronutrient therapy should be a cornerstone of both prevention and treatment.

Bioavailability

As we don't want to fill our medicine cabinet with an indiscriminate selection of potions and pills, we first need to consider the dosage form, absorption, and bioavailability of certain dietary supplements.

The bioavailability of a supplement is a measure of how quickly it's taken up by the body and delivered to the right place. Suppose your thyroid gland is not functioning properly. The cells of your thyroid send out signals to say your body is deficient in iodine, and so you take an oral supplement. It enters your system through your stomach, which delivers the iodine into your bloodstream. The iodine is then pumped round your body until it ends up where it needs to be, thanks to receptors on the cells of the thyroid gland.

The supplement is then effectively present in the target cells, but it's not necessarily active yet. The bioactivity of a supplement depends on the form it's taken in. There are natural and synthetic versions of minerals and vitamins, and a commercial package might include the active or inactive form of the substance. We are creatures of nature and so I will advocate as a principle for as natural an approach as possible. In plain terms, this means you should aim to get as many nutrients as possible from the food you eat and only use supplements to fine-tune your otherwise-perfect diet.

When you do take supplements, opt for quality, and choose natural and bioactive forms. It may sometimes be the more expensive option, but a substance in its natural form often has the advantage of being far more active within your body. Some supplements are sold in inactive forms — the coenzyme Q10 is a good example of this. Taking it in such a form isn't optimal; take the active form of Q10, and it will be utilised far better in your body's cellular processes.

How much of a given substance do you need, and do you have the right enzymes to convert it? Once again, there's no one-size-fits-all solution.

The pharmacological form of a supplement is also important: certain supplements are better absorbed when taken in powder form; others are better in capsules or slow-release tablets.

When nutrients are compressed to make a pill, they have to be mixed with filler substances and binding agents. This sometimes affects the solubility of the nutrients in unpredictable ways. You can test the quality of supplement tablets yourself. They should disintegrate fairly quickly when you drop them into a glass of water. If your tablet is still happily sitting in the water without dissolving after three days, it will likely do the same inside your body. Micronutrients in capsule form don't require as many fillers and binders, and their absorption rate is less dependent on digestion.

Just Pop a Pill and That's All You Need?

Let's now take a look at dietary supplements from the scientific perspective. We know a great deal about the benefits of a number of vitamins and minerals. A sufficient intake of vitamin D, for instance, lowers your risk of respiratory infection. We know vitamin C speeds recovery from illness, although it's very difficult to determine the best dosage or form you should take. That's why it's important not to choose just any old vitamin-C preparation, but to take it in a form with high bioavailability and bioactivity. But what exactly happens when you pop a vitamin C pill?

Vitamin C is water-soluble. It's processed by your digestive system and ends up in your bloodstream. However, the proportion that's absorbed by your body is limited, and not all the vitamin in your digestive system will make it into your bloodstream. Some of it will be excreted in your stool. Some of the portion that does end up in your blood passes through your kidneys and is excreted via that route. So, of the pill that you swallowed, only a tiny proportion ends up where it needs to be. And even that amount differs from person to person. As I said before: one size fits all might be fine for woolly hats, but certainly not for dietary supplements.

In the United States, it has been customary for some time to administer micronutrients intravenously through a drip. If you compare the price of a pack of supplement pills with the cost of an infusion, the latter appears ridiculously expensive. Yet an increasing number of sources confirm that there is indeed an added value to this form of dietary supplementation. The absorption of many products when taken orally is quite limited, and is often made worse by smoking, stress, other medications, and the quality of a person's gut wall. Injecting nutrients straight into the bloodstream means the active substances can reach the place where they're needed in higher concentrations.

Anyway, moving on from this short introduction to pharmacology, let's take a look at a few micronutrients in more detail.

Micronutrients

Zinc

Zinc is a trace mineral that helps our immune cells develop and stay healthy. Over the past 50 years, an abundance of scientific evidence has accumulated showing zinc's antiviral properties through various mechanisms. On the basis of scientific research, zinc is used in the treatment of infections caused by the herpes and common-cold viruses, for example. So, there is a great deal that we know, but, as always, there is just as much that we don't.

The higher the zinc level, the less chance viruses have to multiply; this mineral is effective in hindering virus reproduction. And, although viruses do need a tiny amount of zinc to survive, they are unable to tolerate high levels, and die as a result.

Most people get enough zinc through their diet. Basil, salmon, prawns, mushrooms, garlic, almonds, and red meat are among the foods that are rich in zinc. It's not yet generally recommended that patients start taking gigantic amounts of zinc when they have an infection, except in the case of acute viral infections such as a cold, and then usually in combination with vitamin C. In principle, there's no need to take zinc supplements, except for people who are unable to store sufficient zinc from their diet, as a result of intestinal problems, for example. Zinc supplements are best taken in the form of zinc picolinate, which is the most bioavailable and bioactive form, meaning it's the form that's absorbed most rapidly through the gut and transported into the bloodstream. The recommended daily dosage for zinc picolinate supplements is 15 to 30 grams.

Zinc is also used to treat or prevent zinc deficiency and its consequences, which include stunted growth and acute diarrhoea in children, and delayed wound healing. Zinc is also used in the treatment of Wilson's disease — a condition that causes excess copper to build up in the body. A vegetarian or vegan diet can increase a person's zinc requirement by up to 50 per cent, and older people also require a higher intake of zinc. Proteins, garlic, and onions increase the body's ability to absorb zinc.

Copper

Copper is also a trace mineral that plays an important part in keeping your immune system healthy and in helping your body to respond adequately to antioxidants. Keeping your copper intake at the right level will support your immunity and prevention of infections. However, copper is a nutrient we should handle with care: both too much and too little copper can actually promote infections.

Most people don't need to supplement their copper intake, since the element is present in sufficient quantities in their diet. But let me make an important note on the side here: there must always be a balance in your body between copper and zinc. If you start taking crazy amounts of copper supplement, you will affect your body's ability to absorb zinc. Not only is the equilibrium lost, but it will also cause a state of inflammation in your body. Conversely, if you're taking zinc supplements, you should be sure to eat enough food that is rich in copper, such as liver and seafood like oysters, crab, or squid, as well as dark leafy greens, avocados, nuts, seeds, beans, chickpeas, and dark chocolate.

But how can you tell if those two minerals are in balance in your body? Measure to manage, dear reader, measure to manage. Don't just take any old stuff at random every now and then. I don't often prescribe copper supplements for my patients, except when I can see that their zinc–copper equilibrium is badly out of kilter.

Copper plays an important part in the transport and absorption of iron and can be used in the treatment of anaemia, osteoporosis, premature hair greying, and neurological growth retardation. Excess copper intake can lead to gastro-intestinal problems, liver damage, kidney failure, and even coma. Copper enhances the effect of anti-inflammatories, which can increase the risk of stomach ulcers.

Vitamin D

I went into great detail about this vitamin in the previous chapter. The primary source of vitamin D is sunlight, but, in regions where UV light is weak (or, conversely, where the UV light is strong and the population needs to protect themselves from it), it's advisable to take extra vitamin D as a supplement. More than 50 per cent of the world's population suffers from a vitamin D deficiency — mostly without even realising it!

Vitamin D is important both for bone health and to strengthen the body's immune response. I already mentioned that the recommended blood level of vitamin D is at least 30 nmol/L, but the correct amount is more likely to be 50, or even 60, nmol/L. Taking vitamin D as a dietary supplement is usually advisable in most cases. It is a fat-soluble vitamin, so is best taken with a fat-rich meal, otherwise it can't be absorbed as well by the body.

Conditions associated with a lack of adequate vitamin D can vary greatly. Examples affecting the musculoskeletal system include osteoporosis and bone fractures. Inadequate vitamin D levels can also play a part in various autoimmune diseases, such as type 1 diabetes, multiple sclerosis, Hashimoto's disease, and obesity. To a certain extent, cardiovascular disease can also be related to a vitamin-D deficiency. Various medications can negatively affect vitamin-D levels, and so require extra monitoring of the levels or extra intake of vitamin D.

NAC

N-acetyl-L-cysteine is an amino acid that plays a very important role in the production of glutathione, which is your body's master antioxidant. It helps to reduce, and keep to a minimum, the oxidative stress in your cells caused by the free radicals released when oxygen is consumed. NAC can be used to reduce inflammation and mucus production in people with certain lung diseases. COVID-19 also causes a build-up of fluid or mucus in the alveoli of the lungs. That impairs their ability to function and allows oxidants to accumulate there. That's when the situation turns into an emergency. In order to be able to produce large amounts of glutathione, your body needs cysteine, otherwise it can't start the antioxidant process.

'Ordinary' cysteine is found in high-protein foods such as chicken, eggs, fish, garlic, onion, broccoli, and other vegetables from the brassica family. The specific kind of cysteine you can take to increase your body's production of glutathione and boost your antioxidant system — NAC, which is actually far more bioactive — exists only in synthetic form.

Paracetamol Poisoning

Few people realise that paracetamol, that common painkiller, is actually also a silent killer. It's freely available over the counter, and the maximum daily dosage is four grams (eight tablets). Got a headache? Take a paracetamol. Feeling feverish? Paracetamol. It can't do any harm, right? Wrong! Just ten grams a day can cause serious liver poisoning, even leading to sudden liver failure. Paracetamol poisoning is more common than you'd think, and the symptoms often don't appear until 48 to 72 hours later. In hospitals, NAC is used as an antidote to paracetamol poisoning. (NAC can help even with an ordinary hangover, but don't tell anyone I told you.)

Vitamin C

I've talked a lot about vitamin C throughout this book. And it's a super-interesting and super-important vitamin. Vitamin C is an essential nutrient that's important in the production of collagen. Insufficient vitamin C causes scurvy.

Vitamin C is also a powerful antioxidant that works at various levels. Does it make sense to take it for a sore throat? Certainly. It supports the cells of your frontline defence, such as in your mucous membranes. Vitamin C also offers vital support to your body's antiviral response. When your immune cells spring into action due to a viral infection, the oxidative stress drives them crazy and your vitamin C levels plummet. Taking extra vitamins during an acute infection thus makes a lot of sense, even just to maintain normal levels.

Vitamin C is also a natural antihistamine, albeit to a lesser extent. Do you remember from our immunology course what the function of histamines is? When histamines are released, the blood vessels dilate so that inflammatory substances can more easily reach the part of the body in question. If that's in your nasal cavity, it will cause sneezing fits or a runny nose. Taking vitamin C will help reduce the swelling locally.

What's the recommended daily amount? It won't surprise you to read that there's a lot of debate about this. 110 mg a day should be an appropriate dose. But that doesn't count for smokers. Inhaling smoke creates massive amounts of oxidative stress and so the amount should be increased immediately.

In the case of acute illness, you can take even higher doses — something known as a vitamin-C flush. As I described earlier, we have no idea how much vitamin C effectively remains in the body and is absorbed. A vitamin-C flush usually involves an increase of 1.5 to 2 grams a day — far more than the maximum recommended daily intake! However, some people have difficulty dealing with such a high dose and may experience side effects such as diarrhoea. Vitamin C is one of the micronutrients that can be administered intravenously.

Vitamin E

Vitamin E plays an important part in keeping your heart, brain, immune system, and skin healthy, as well as being important for fertility. It's a fat-soluble vitamin, but it isn't advisable to take it willy-nilly. In excessively high doses, vitamin E does more harm than good, affecting the blood's ability to clot, for one thing. In addition, the form the vitamin takes in synthetic preparations is often not the form your body needs and is able to process. There's little evidence that healthy people benefit from taking vitamin-E supplements.

You can get enough vitamin E from natural sources, such

as eggs, nuts, spinach, broccoli, shrimp, onions, sunflower seeds, and peanuts.

Clinically, vitamin E may have an immune-boosting effect and protect against cardiovascular disease and certain types of cancer. An increased vitamin-E requirement can occur in people who are unable to absorb fats very well, in endurance athletes, and in those with metabolic syndrome.

Vitamin B

I could write a whole book just about vitamin B! It's a water-soluble vitamin — or rather vitamins, as there are different kinds, which carry out very different tasks in your body. Some of those tasks are key to the correct functioning of your immune system. The B vitamins are numbered, starting at B_1 and ending at B_x — when the numbers run out, they are replaced by letters. However, there is not an equal amount of scientific evidence regarding every kind of vitamin B.

How do B vitamins work? To simply list the facts: they increase the number of T lymphocytes, they increase the activity of macrophages, they help the immune system to recognise bacteria, and they increase the activity of natural killer cells. In view of this, it makes sense to take a complex of different vitamins to support your immune system. A regular multivitamin preparation containing a range of B vitamins can't generally do much harm, since B vitamins are water-soluble: that means any excess you take in will simply be excreted in your urine.

The B-vitamin levels most likely to be checked by your doctor are those of B_{12}, B_9, and B_6. It's best not to have too much or too little of vitamin B_6, as an excess can cause a nerve-tingling sensation, leading to neuropathy. Unfortunately, this is a common problem among athletes, as they tend to consume so many supplements and sports drinks. B vitamins that are checked less frequently, but which still have significant effects on our body, are B_1, B_2, B_3, and B_5.

Vitamin B_1 (thiamine) plays a role in the production of energy from carbohydrates. It's non-toxic, and deficiencies are associated with chronic alcoholism or a low dietary intake due to anorexia, excessive vomiting, or old age. Pregnancy, stress, or hard physical work can also result in a higher vitamin-B_1 requirement.

Vitamin B_2 (riboflavin) is an essential, water-soluble vitamin. It is critical for the conversion of vitamin B_9, vitamin B_6, and tryptophan, and the absorption of iron, zinc, and calcium. It may play a part in migraine prevention.

Vitamin B_3 (niacin) is important in our energy metabolism and blood circulation, as well as in helping us to maintain a healthy blood-cholesterol level.

Vitamin B_5 (pantothenic acid) protects our cells against oxidant damage by increasing glutathione production.

Vitamin B_6 is involved in the regulation of our homocysteine levels. Too much homocysteine remaining in the bloodstream can damage the artery walls. A vitamin-B_6 deficiency can cause the level of vitamin C in our blood to fall, while too much of it can lower the amount of glucose in the blood and lead to a drop in blood pressure. And, as mentioned above, large doses of vitamin B_6 can cause nerve damage.

Vitamin B_9 (folic acid) is essential for the production

of DNA/RNA, for cell division, and for the development of the fetal nervous system. Folic-acid supplements can help people with prolonged diarrhoea and infections. Folic acid is given to pregnant women to reduce their baby's risk of birth defects. It's important to monitor the vitamin-B_{12} status of anyone taking folic-acid supplements. Many medications have a negative effect on folic-acid levels, so in such cases it's best to have it checked.

The bioactive form of vitamin B_{12} is called methylcobalamin. It contains cobalt and is produced, to a limited extent, by microbes in our gut. It's also present in animal-based foods. Vegetarians and vegans often have a vitamin-B_{12} deficiency and may benefit from taking supplements. Vitamin B_{12} plays a role in the development and growth of cells and is essential in removing homocysteine from the blood. B_{12} is often called the body's natural energiser, because it's a multipurpose vitamin. Many people have a normal but less-than-optimal level of vitamin B_{12}. An inefficient gut can lower the absorption rate of this vitamin, but some people actually lack a certain enzyme (gastric intrinsic factor) and are unable to absorb it at all. These people have to get their vitamin B_{12} via a different route, usually intravenously or by injecting.

Iron
Your body needs iron to maintain an efficient immune response, to resist infection, to produce red blood cells, and to transport oxygen around. It also protects your cells against possible damage from free radicals. If you're struggling with an iron deficiency, you will feel tired, you will be unable to

take up enough oxygen, and your immune response will be less effective. But too much iron is also less than ideal: viruses and bacteria feed on excess iron. The last thing you should to do when you have an acute infection is take iron supplements! Blood tests often include an iron-content check to see if your red and white blood cells are of sufficient quality.

You can get sufficient iron largely from your diet, and, no, you don't have to eat giga-amounts of spinach! The whole Popeye story is based on research undertaken in the 19th century in which the scientists misplaced a decimal point when recording their results, instantaneously attributing a much higher iron content to spinach than it actually has. Good dietary sources of iron include nuts, seeds, avocado, beans, and apricots, among many others.

Lactoferrin

What you *can* take while you have an acute infection is lactoferrin. Produced by various glands and in white blood cells during an infection, this glycoprotein helps regulate the immune system somewhat: it pumps up your natural killer cells and cytokines, activates your B and T lymphocytes, and is also great at killing agents of infection — making it an interesting dietary supplement. When it binds to iron, it renders the metal inedible to bacteria, so helping to stop them growing and spreading.

Selenium

Selenium is a crucial antioxidant. This mineral also specifically supports the body's antiviral response. A selenium deficiency —

a common condition among the elderly — weakens the body's immune response in general, and even otherwise-harmless viral infections can become devastating, or fatal. There's much discussion about whether taking selenium supplements is a good idea, and especially about the recommended daily amount. In my opinion, taking selenium isn't necessary, as it's easy to cover your selenium requirements with a healthy and varied diet. A selenium deficiency despite a healthy diet may be an indication of a genetic defect.

Herbs and Foodstuffs

There should be more inside your medicine cabinet than just micronutrients. That's why I'd like to review the medical effect of certain herbs and foodstuffs.

Omega-3

I know omega-3 is neither a food nor a herb, but I think this macronutrient deserves a place in the medicine cabinet. It's simply too important a substance to leave out! The best thing you can do to make sure every cell membrane in your body is supplied with a decent amount of omega-3 fatty acid is to regularly include some salmon, nuts, seeds, avocados, and olives in your diet.

Omega-3 fatty acids are important in maintaining the balance between promoting and preventing inflammation. In general, the omega-3 fatty acids have an anti-inflammatory effect and the omega-6 fatty acids are pro-inflammatory. The ratio between these two families of fatty acids should be in omega-3's favour. This lowers the concentration of unhealthy fats in the blood and prevents the platelets in the blood from sticking together and causing clotting. In addition, omega-3 reduces our internal production of arachidonic acid, a pro-inflammatory form of omega-6.

People take omega-3 fatty-acid supplements as a general support for their health, to curb so-called inflammaging ('inflammation' plus 'ageing' — the chronic low-grade inflammation that develops with advanced age), and to improve the balance between omega-6 and omega-3 fatty acids. Such supplements can also provide protection against stomach ulcers and increase the effectiveness of probiotics. The possible side effects include gastrointestinal complaints. Not infrequently, people who take these supplements suffer from nausea or fishy burps. Omega-3 fatty-acid supplements are not recommended for people with a hypersensitivity to fish (oil), algae, or krill (oil).

Garlic

Garlic is in the same family of plants as the onion. I love it, though I know people's opinions on garlic vary widely — especially when it comes to the smell you disseminate after eating it. Garlic has very few calories but a high nutritional value. Raw garlic in particular is rich in natural, active substances and nutrients, such as manganese, vitamin B_6, vitamin C, and selenium. Garlic strengthens the immune response, reduces inflammation, and has a positive influence on cardiovascular problems, as well as diabetes. In a study in which test subjects took garlic for 12 weeks, they were found to have a 60 per cent lower incidence of the common cold than members of a control group who didn't take it. Garlic actively lowers blood pressure and has a positive effect on blood-cholesterol levels. Many of the substances in garlic help capture free radicals. Olympic athletes in Ancient Greece used garlic as a doping agent, and recent studies have shown that it does indeed improve exercise performance. It's also good for bone health and helps detoxify the body.

How should you take garlic? The best option is simply to eat it raw — or include it often in your cooking. But it is also available in capsule form.

Echinacea

A lot has been written and said about echinacea, so I have to be careful about making definitive statements concerning this plant. It seems that for every study on echinacea there is a counter-study. But echinacea, commonly called the coneflower, is said to be a natural immune booster. Preparations made from echinacea may render viruses partially inactive — especially those that produce mild, cold-like symptoms — by destroying the viruses' outer membranes and their protein coat. It doesn't prevent the actual infection, but it is thought to affect the duration and severity, by reducing the overstimulation of mucus production in the respiratory tracts.

When taking echinacea preventively, it's best to take a course for ten to 15 days a month, to avoid habituation of your immune system.

Resveratrol

Resveratrol is a polyphenol found in, among other things, blueberries, dark chocolate, and peanuts, and particularly in the skin of black grapes. It's the antioxidant that makes red wine 'healthy'. It has a double antioxidant effect: on the one hand, it captures the free radicals released when we use oxygen; on the other, it supports all the enzymes of the antioxidant system. So, it's all good news, then ...

Are you waiting for me to advise you to drink more wine for your health now? A glass of red wine a day will not do you any harm. However, there is a major drawback to resveratrol: it has very low bioavailability and is therefore absorbed by the body only to a moderate extent. *Aha,* thinks the alert reader, *I'll just drink more wine then!* But, no, that's not how it works. Alcohol is sugar and increases your insulin response, so drinking more alcohol will do you more harm than good. There is something you can take to increase your body's absorption of resveratrol: piperine — the alkaloid responsible for the pungency of black pepper. A good twist of the pepper mill into your glass of red wine will instantly increase the bioavailability of the resveratrol it contains. ;-)

Shiitake

This edible mushroom is the pearl of Asian traditional medicine. It is widely acclaimed for its nutritional and medicinal properties. The shiitake mushroom is used in its dried form to boost the immune system, to increase cardiovascular and liver health, and to support the immune cells in the case of infections like the flu, or even hepatitis. The recommended dose of a powdered shiitake preparation is around 15 grams per day.

Oregano

Oregano is often used as an antiseptic in natural medicine. It supports your cardiac health and antioxidant defences.

It's also said to have an active effect on the protein coats of viruses, and the phenols thymol and carvacrol found in oregano can protect against viruses that cause diarrhoea and vomiting.

Licorice

Licorice (or liquorice), extracted from the root of the *Glycyrrhiza glabra* plant, is used in natural medicine to treat digestive problems, coughs, menopausal disorders, and infections. It is also said to have antiviral properties and to trigger various mechanisms that support T lymphocytes, as well as stifling the gene expression of viruses — that's when a virus inserts its DNA into the DNA of a cell. People who struggle with high blood pressure should proceed with caution, as blood pressure can be raised by licorice.

Honey

Delving deeper into grandmother's cabinet of traditional cures, we find that raw honey is chock-full of nigero-oligosaccharides. Woo hoo! As a defender against respiratory infections and a panacea against dry coughs, this sugar bomb with antimicrobial properties and immunity-boosting effects deserves a prominent place in any self-respecting household's medicine cabinet. With its very sticky consistency, it acts as extra protection against bacteria trying to penetrate your frontline defences.

But, as they say, every silver lining has its cloud. Honey provides such a huge fructose shot that it causes insulin levels to peak sharply and can even trigger diabetes in the long term. And, due to its high sugar content, it's also ideal fodder for bacteria, enabling them to divide and multiply all the more quickly.

So, should you be eating honey or not? Of course you should — but in moderation! A dollop of honey in your tea, a spoonful in salad dressings or marinades, a thin layer on a slice of spelt or wholegrain toast with a little bit of peanut or almond butter ... delicious, but best not just before bedtime, as we've no need of an insulin boost when we're about to rest.

Bee Pollen, Propolis, and Royal Jelly

These bee products are used in traditional medicine for their immune-boosting properties.

Bee Pollen

This dietary supplement is used to strengthen the immune system, fight microbial infections, and keep the liver healthy. It's rich in immunostimulatory polysaccharides — in particular, glucans and galactans. Pure bee pollen as a supplement mostly contains beta-glucans (or bee-ta-glucans ... I'm sorry), which are known to bee (I just can't resist these puns) particularly good at preventing cancer and boosting the immune system.

Propolis

Propolis is the stuff that bees use to seal up unwanted holes in their hive. Propolis extract has a positive effect on the chronic symptoms of a number of respiratory infections.

Royal Jelly

This milky substance, which is primarily intended as food for the new queen in a bee colony, stimulates the production of T cells. It also has masses of other anti-allergenic properties!

Ginger

Packed full of antiviral and antimicrobial properties, this super-healthy spice has a beneficial effect on the immune system's third-line defences. Ginger also positively affects the digestive system and is now increasingly used as a remedy for nausea.

Elderflower/Elderberry

This supplement is often taken (in the form of syrup, for example) when flu symptoms appear. It's said to be beneficial in treating fever, headaches, muscle pain, and congestion. It makes less sense to take elderflower preventively.

Ashwagandha

This medicinal herb is the great supporter of lung function. It has a slightly antiviral effect on viruses such as herpes, hepatitis, and HIV. It also has an effect on the adrenal gland and helps stabilise stress-hormone levels.

Curcumin

Curcumin — present in the spice turmeric — creeps indirectly into cell nuclei, where it acts on gene expression just like a virus and lowers the cell's capacity to produce inflammatory substances. Its anti-inflammatory effect is powerful! This substance is best taken as a supplement, as you would have to ingest far too much powdered turmeric to get the same effect (while having a less than positive effect on the taste of your food). Plus, the spice itself is not absorbed very well by the body, unless you add a little pepper to it.

To spare you any culinary calamities, rather than suggesting what you should cook myself, I asked health coach Anouk Brackenier to come up with some relevant recipes. And she has succeeded in conjuring up some delicious recipes using ingredients from the traditional medicine cabinet.

'Good health
starts in the gut'

Hippocrates

5
RECIPES

Hello everyone!

When Servaas told me he was planning to write this book, and asked me to contribute some recipes, my reaction was: 'Hell, yeah!' My assignment was to provide recipes that boost your immune system. In times of the coronavirus, it's more important than ever for all of us to do whatever we can to avoid illness for ourselves and others, to take care of our bodies as well as we can, and to strengthen our resistance so that potentially harmful microbes can't get the better of us. As you've already read in this book — and, as a health coach, I can't stress this enough — it all depends on the positive interaction of several strands: diet is not the only important aspect here; exercise, stress management, sufficient sleep, and a positive mindset are all important elements in feeling good and staying healthy.

Servaas and I share a passion for making (and keeping) people healthy through lifestyle changes! But let's be honest, diet and nutrition have always been and will always remain a mainstay of that process. That's part of the reason why I was over the moon to get in the kitchen to seek out great recipes and pursue that passion we share.

And, to be honest, I have enough ideas to fill dozens more of Servaas's books ... I can't wait!

I hope you enjoy these recipes, and that you take good care of yourselves and others. And: keep on sharing — both the love and the recipes! :)

You only have one body and one mind, so take care to keep them healthy.

Love,
Anouk
Insta: @health_coach_anouk

VITAMINS AND MINERALS

ZINC

Soba Noodles with Scampi and Navy Beans

The zinc-rich stars of this dish are the scampi, spinach, and sesame seeds. The lime juice and fresh parsley add extra vitamin C for free! Real soba noodles are made of 100 per cent buckwheat and so are gluten free — ideal for people with an allergy or intolerance to gluten. But be aware that some varieties are made from a combination of buckwheat and regular wheat! Remember to wash the rind of the lime thoroughly, and choose organically grown limes if you can.

INGREDIENTS (SERVES 4)
- Olive oil
- 3 cloves garlic, crushed
- 1 onion, finely chopped
- Salt and pepper
- Cayenne pepper (optional)
- 4 tomatoes, diced (or one 400 g / 14 oz tin of chopped tomatoes)
- 100 g (3½ oz) baby spinach
- 200 g (7 oz) canned navy/haricot beans (drained weight), rinsed thoroughly
- 3 tbsp chicken stock
- 3 tbsp dry white wine
- 1 handful of fresh basil leaves
- 1 tsp lime zest
- Juice of ½ a lime

- **240 g (8½ oz) soba noodles**
- **16 large scampis, cleaned and gutted (thawed if using frozen scampis)**
- **A few tbsp sesame seeds**
- **Fresh parsley**

METHOD

- Heat a little olive oil in a large pot and fry the garlic and onion. Season with a little salt and pepper. Add a pinch of cayenne pepper if you like it spicy!
- Add the chopped tomatoes and sauté for 2 to 3 minutes. Next, add the spinach and stir until the leaves have wilted. Add the (drained) beans, chicken stock, and white wine, and simmer for a few minutes until only a little liquid remains. Don't let it boil dry.
- Add the basil, lime zest, and lime juice, and mix well. Season again if necessary and keep warm.
- Cook the soba noodles in salted water as instructed on the packet.
- Pat the scampis dry. Fry them briefly on each side in hot olive oil on a high flame until they turn pink. It should take just 1 to 2 minutes.
- Put the soba noodles in a dish or bowl, pour over the bean mixture, and put some of the scampi on top of each dish. Garnish with the sesame seeds and freshly chopped parsley.

Chicken with Spicy Peanut Sauce

The zinc suppliers in this dish are the chicken, asparagus, and mushrooms, with extra support from turmeric, ginger, and garlic.

I admit it: I'm a real peanut-butter lover! I could unashamedly wolf down a full jar of the stuff by scooping it out with my bare fingers ... So, #sorrynotsorry that several of these recipes are enriched with this lovely nutty treat. :) And peanut butter is also full to bursting with healthy omega-3s! Check the information on the jar to make sure it doesn't contain added sugar — there are lots of different brands and types available! For anyone with a peanut allergy, the peanut butter can be replaced with any other nut paste.

INGREDIENTS (SERVES 4)
- 800 g (28 oz) chicken fillets
- Olive oil
- Salt and pepper
- 200 g (7 oz) green asparagus tips
- 250 g (9 oz) button mushrooms, halved

MARINADE
- 1 tsp fresh ginger, grated
- 1 tsp turmeric
- 1 tsp powdered coriander (aka cilantro; Servaas leaves this out)
- 4 tbsp melted coconut oil
- Pepper, a few good twists of the mill
- 1 tsp curry powder
- ½ tsp salt

PEANUT SAUCE

- 1 clove garlic, finely chopped
- ¼ of a red chilli, seeded and finely chopped (feel free to use more if you like it extra-spicy)
- Olive oil
- 60 g (2 oz) peanut butter
- 200 mL (7 fl oz) coconut milk
- 1 tbsp soy sauce

METHOD

- Dice the chicken fillets. For the marinade, mix the ginger with the turmeric, coriander, and curry powder, along with the coconut oil and the salt. Add a good twist of freshly milled pepper. Pour the marinade over the diced chicken, mix well, and leave to marinate for at least 10 to 15 minutes.

- Make the peanut sauce: in a saucepan, fry the garlic with the chilli in a little olive oil to release the flavour. Add the peanut butter and stir well until it is 'melted'. Gradually add the coconut milk, bring it to the boil, and then turn down the heat and simmer for 5 to 10 minutes, until the sauce thickens.

- In a hot wok or frying pan, fry the diced chicken rapidly in some olive oil. Reduce the heat to medium and continue frying the chicken until cooked through. Season to taste with salt and pepper.

- Remove the diced chicken from the pan and put to one side (keeping it warm). In the same pan, fry the asparagus tips and mushrooms in coconut oil or olive oil until cooked through.

- Serve the chicken with the peanut sauce and fried vegetables. Bon appétit!

VITAMIN D

Since there's a better than even chance you don't get all the vitamin D you need from sunshine, these recipes will give you an extra boost!

Tuna Muffins

The vitamin D in this dish comes from the tuna and the egg yolk. One yolk contains more than half your daily vitamin-D requirement. Remember what Servaas said about daily intake — moderation in all things! These muffins have additional nutritional benefits from the garlic and the vitamin C in the bell peppers.

INGREDIENTS (MAKES 12 MUFFINS)
- 120 g (4 oz) bell peppers (capsicums)
- 100 g (3½ oz) green olives
- 1 shallot, chopped
- 300 g (10 oz) tuna (canned in its own juice, drained weight)
- 3 eggs
- 100 g (3½ oz) almond yoghurt (or other yoghurt of your choice)
- 2 tsp ras el hanout
- Pepper and salt
- 1 clove garlic, crushed

YOU WILL ALSO NEED
- (Silicon) muffin cases

METHOD

- Preheat the oven to 180 °C (350 °F).
- Wash the bell peppers and dice them finely. Slice the olives and chop the shallot. Drain the tuna and break up any larger pieces.
- Beat the eggs and stir in the yoghurt. Add the bell peppers, shallot, (crushed) garlic, olives, and tuna, and mix well. Season with ras el hanout, pepper, and salt. Put an equal amount of the mixture in each muffin case (the cases should be about ¾ full). Bake in the oven for about 15 minutes. The muffins are done if a skewer comes out clean of any 'dough' when you prick them.
- Serve with a salad or a nice bowl of soup.

Sardine Tapenade

Sardines are a fantastic source of vitamin D and omega-3, the fatty acid that protects against cardiovascular disease and has anti-inflammatory properties. Sardines are also a source of calcium, selenium, potassium, magnesium, and vitamin B_{12}. Those last two are crucial in energy production. Sardines also contain many proteins that are important for the production of antibodies. Extra fact: sardines feed exclusively on plankton and contain a lower concentration of mercury than other kinds of fish. Lemon and parsley add the perfect finish to this dish with a good dose of vitamin C.

INGREDIENTS (SERVES 4)

- 180 g (6 oz) sardines (canned, drained weight)
- 2 tbsp lemon juice
- 1 tbsp capers

- 1 handful of parsley
- 50 g (2 oz) green olives, pitted
- 3 tbsp olive oil
- 1 clove of garlic
- 1 tbsp mustard
- Salt and pepper

METHOD

- Drain the sardines. Put them and all the other ingredients in a food processor, and mix to a smooth tapenade. Season with pepper (and salt). Delicious on gluten-free toast or in a sandwich.

SELENIUM

Cod with Asparagus and Brazil-Nut Pesto

Asparagus is heaven for the little creatures in your belly. Together with the brazil nuts and the cod, it provides a mega-hit of selenium. Brazil nuts especially are a real 'selenium bomb'! Raw garlic is a good detoxifying agent due to the sulphur compounds it contains.

INGREDIENTS (SERVES 4)

- 4 cod fillets
- 2 bunches of green asparagus
- Black pepper and salt

BRAZIL NUT PESTO

- 120 g (4 oz) brazil nuts

- 50 g (2 oz) pine nuts
- 1 avocado
- 2 large handfuls of rocket (arugula) salad (about 50 g / 2 oz)
- 1 handful of fresh parsley
- 1 handful of fresh basil
- Juice of 1 lemon
- 8 tbsp of olive oil
- 1 clove of garlic
- Salt and pepper

METHOD

- Preheat the oven to 180 °C (350 °F).
- To make the pesto, finely chop the brazil nuts and pine nuts in a small blender or food processor. Add the avocado, rocket, parsley, basil, lemon juice, olive oil, and garlic, and season to taste with salt and pepper. Mix to form a smooth, creamy pesto.
- Season the cod fillets on both sides with salt and pepper. Spread the top of each fillet with a layer of pesto and place them side by side in a baking dish. Bake in the oven for about 15 minutes.
- In the meantime, blanch the asparagus in salt water, drain, and add it to the cod in the oven for the last 5 minutes.

Tip: You can use any leftover pesto in the immunity burger!

Colourful Tofu Curry

The selenium sources in this dish are the tofu, cashews, navy beans, and brown rice. The beans and rice will be welcomed as a feast by your microflora.

INGREDIENTS (SERVES 4)

- 1 zucchini (courgette)
- 1 red pepper
- 1 onion
- 1 clove of garlic
- 1 tsp of fresh ginger, grated
- Olive oil
- 400 g (14 oz) tofu, diced
- 100 g (3½ oz) baby sweet corn
- Salt and pepper
- 2 tbsp curry powder
- 1 tsp turmeric
- Brown rice
- 250 mL (8½ fl oz) coconut milk
- 200 g (7 oz) navy/haricot beans or cannellini/white kidney beans (drained weight, canned, well rinsed)
- 1 handful of cashews
- 1 handful of fresh herbs (coriander/cilantro or parsley)

METHOD

- Wash the zucchini and the red pepper and roughly dice them. Chop the onion. Peel and crush the garlic clove. Peel and grate the ginger.
- In a hot pan, fry the onion together with the garlic and fresh ginger in olive oil to release the flavour. Add the tofu, zucchini, red pepper, and baby sweet corn. Season with salt and pepper, curry powder, and turmeric, and cook for about 5 minutes.
- Meanwhile, cook the brown rice according to the instructions on the packet.

- Pour the coconut milk into the tofu and vegetable mixture, and add the beans. Leave to simmer (uncovered) on a medium heat for 10 minutes. Serve with the brown rice, and garnish with a handful of cashews and fresh herbs of your choice.

VITAMIN C

Vitamin C is an important ally in the fight against viruses. It stimulates your immune system and helps keep your defences up. It's found in all kinds of fruit and vegetables. Various kinds of cabbage, as well as peppers, citrus, kiwifruit, parsley, and strawberries all provide a real vitamin-C boost.

C-Booster Smoothie

Smoothies are a quick and easy way to get all the vitamins you need in one go. So, here we go! This particular smoothie is spinach-, kiwi-, and mango-based. That's a vitamin-C triple whammy!

INGREDIENTS (SERVES 2)
- 1 kiwifruit
- ½ a mango
- 1 handful of spinach leaves
- Juice of ½ an orange
- 180 mL (6 fl oz) plant-based milk of your choice

METHOD
- Put all the ingredients in a blender and whizz!

Funky-C Froyo Lollies

Frozen yoghurt (or froyo) is an excellent way to enjoy ice cream without the sky-high sugar peak. On the contrary, in fact, this treat even gives your health a nice boost. Get those vitamin-C levels up! Melon, kiwifruit, and lime are the C-stars here, providing your white blood cells with the necessary vitamin C.

INGREDIENTS (SERVES 4)

- 225 g (8 oz) melon (cantaloupe)
- 3 kiwifruits
- 350 mL (12 fl oz) coconut yoghurt
- Juice of ½ a lime
- 100 g (3½ oz) coarse coconut flakes

METHOD

- Put all the ingredients (except the coconut flakes) in a blender, and whizz to a smooth paste. Add the coconut flakes, and whizz briefly again until well mixed.
- Pour the mixture into popsicle moulds and leave in the freezer for a couple of hours until solid.

Tip: In summer, also add 200 g (7 oz) of strawberries to the mix. They are chock-full of vitamin C.

Vegetable Bake with Bacon Bits

Fruit isn't the only source of vitamin C: vegetables are full of it, too. Bell peppers, for example, could easily be renamed C-peppers, they're so rich in the vitamin. This vegetable bowl is also bursting with vitamin D, and broccoli and brussels sprouts have a number of other antioxidant properties, too.

INGREDIENTS (SERVES 2)

- 600 g (21 oz) brussels sprouts
- 2 sweet pointed peppers
- 1 head of broccoli
- 8 basil leaves
- 2 tbsp pine nuts
- 170 mL (6 fl oz) plant-based cream (soya, oat, or almond cream)
- Salt and pepper
- 1 tsp dried oregano
- 250 g (9 oz) diced bacon bits (or lardoons)
- Olive oil
- 1 handful of grated parmesan cheese

METHOD

- Preheat the oven to 200 °C (400 °F).
- Wash the brussels sprouts, and remove the outer leaves. Roughly chop the peppers. Wash the broccoli, and separate the florets. Whizz the basil and the pine nuts together in a food processor till finely chopped.
- Boil the sprouts and the broccoli (separately) in salted water for 10 minutes (or, better still, steam them).
- Drain the broccoli, and add the plant-based cream, pepper, salt, oregano, and the basil/pine-nut mixture. Mash with a potato masher.
- In the meantime, fry the bacon bits in a (dry) frying pan until slightly crisp. Add the pointed sweet peppers, and fry together for a few minutes, adding a little olive oil

if necessary. Add the brussels sprouts shortly before the end, and fry them briefly with the other ingredients.

■ Put the vegetable-and-bacon mixture in an ovenproof dish, and spread the mashed broccoli over the top.

■ Sprinkle with the parmesan cheese and bake in the oven for 15 to 20 minutes.

Healthy Hutsepot Stew

This healthy version of the traditional Flemish oxtail stew is chock-full of vegetables — ideal for keeping your phyto- and micronutrient levels up! Chinese cabbage, thyme, parsley, and sweet potatoes are almost limitless sources of vitamin C.

INGREDIENTS (SERVES 4)

- 2 leeks
- 3 celery stalks
- 3 parsnips
- 3 carrots
- 200 g (7 oz) Chinese cabbage
- 4 large onions
- 3 medium-sized sweet potatoes
- 2 cloves of garlic
- 1 L (1 qt) chicken or vegetable stock
- 200 mL (7 fl oz) white wine
- 1 tbsp mustard
- Salt and pepper
- 1 tsp dried thyme
- 2 bay leaves
- 1 tsp nutmeg

- **4 thick rashers of bacon**
- **4 sausages**
- **1 handful of parsley**
- **Dijon-style mustard to serve**

METHOD

- Wash and chop the vegetables into rough chunks, about 5 cm (2 in.) in size.
- Stew the vegetables together with the garlic in a large pot for about 15 minutes.
- Pour in the stock, white wine, and mustard, and season with salt, pepper, thyme, bay leaves, and nutmeg. Briefly bring to the boil. Place the rashers of bacon next to each other on top of the vegetables, turn down the heat, and simmer for 1½ hours. Halfway through, turn the bacon over. Don't cover the pot while the stew simmers.
- In the meantime, cook the sausages. Serve the stew with the sausages in a deep plate or soup dish, garnishing with finely chopped parsley and a dollop of mustard.

COPPER

Celery Salad with Walnuts

Earlier in this book, you read that you mainly get the copper your body needs from the food you eat. So, here you go! Sources of dietary copper include chickpeas, rocket, raisins, walnuts, and avocado. How do you fancy a salad made of those ingredients?

INGREDIENTS (SERVES 2)

- 7 celery stalks
- 1½ avocados
- 1 red onion
- 1 apple
- 2 to 3 handfuls of rocket (arugula)
- 280 g (10 oz) chickpeas (drained and rinsed)
- 3 tbsp raisins
- 70 g (2½ oz) walnuts
- 3 tbsp pomegranate seeds
- 4 tbsp olive oil
- 2 tbsp lime juice
- Salt and pepper
- 5 tbsp flaked parmesan
- Fresh parsley (optional)

METHOD

- Wash and finely chop the celery. Keep the celery leaves to garnish the salad later. Dice the avocado, red onion, and apple.
- Put the celery, rocket, red onion, chickpeas, diced apple, raisins, diced avocado, walnuts, and pomegranate seeds into a large salad bowl. Add the olive oil and the lime juice, and mix well. Season to taste with salt and pepper. Garnish with the parmesan flakes and the celery leaves. You can replace the celery leaves with parsley if you prefer.

Sneaky Smoothie

You're going to love this sneaky smoothie: it leaves you feeling full for hours, and you get a portion of zucchini and spinach without you even noticing! This recipe is great for kids, as the vegetables are hidden, and the cacao powder makes it look like a chocolate smoothie! Sneaky but nice! Almond paste provides copper, healthy fats, and a nice, smooth texture. And the honey in this smoothie makes it even more of an immunity boost.

In addition to copper, spinach contains many other minerals, and vitamins, too. It also improves your cognitive skills, strengthens your cardiovascular system, and generally lowers your risk of developing certain medical complaints. So, always add lots of spinach to your smoothies! And don't worry: the rich taste of chocolate means you won't even notice it's there.

Cacao is the purest form of chocolate, as long as we're talking about raw cacao powder (not to be confused with *cocoa* powder, which is made from roasted, not raw, beans and contains fewer nutrients). Cacao's full to the brim with copper, magnesium, and antioxidants. Zucchini is chock-full of vitamin C, and so helps ensure your immune system keeps working properly. It's also full of antioxidants and dietary fibre, and is good for the nervous system.

INGREDIENTS (SERVES 2)
- 1 ripe banana (frozen)
- 350 mL (12 fl oz) unsweetened almond paste
- 100 g (3½ oz) fresh spinach, washed
- 130 g (4½ oz) zucchini/courgette (frozen)
- 2 tbsp unsweetened cacao powder
- 1 tbsp honey (or more if you like it sweeter)
- 2 tbsp peanut butter (or other nut butter of your choice)

In my opinion, peanut butter and chocolate go together as a taste combination like tomatoes and basil, but, hey *chacun à son goût* and all that ...

METHOD
- Put all the ingredients in a blender and 'blitz' them.

Healing Hummus

This dip is a bit 'guacamole meets hummus' ... what more could you want? And it's brimming with omega-3 and protein. It'll make not only your tastebuds, but your whole body happy — and that's the combination we're always aiming for! The copper comes from the avocado, chickpeas, and sesame paste.

INGREDIENTS (SERVES 8)
- 1½ ripe avocados, peeled, stoned, and diced
- 200 g (7 oz) chickpeas
- 15 g (½ oz) fresh coriander/cilantro (or if you don't like that, fresh mint)
- 1 tbsp olive oil
- 2 tbsp tahini (sesame paste)
- Juice of 1½ limes
- 1 clove of garlic
- ½ tsp powdered cumin
- ½ tsp salt
- ¼ tsp black pepper
- ¼ of a red onion
- 1 small tomato, diced

METHOD

- Mix all the ingredients to a creamy consistency in a food processor or blender.

Tip: This dip is great as a starter, but also as a topping on a (low-carb) slice of toast, or as a sandwich filling.

ANTIOXIDANTS

Acai Bowl

Starring role: acai berries
Other antioxidants: blueberries, strawberries
Extra booster in the topping: bee pollen

This acai bowl (if you're not sure how to pronounce it, think 'uh-suyee') is a Brazilian-inspired smoothie bowl with acai berries as the star ingredient. The smoothie is made with frozen mixed fruit and served in a bowl. The deep-purple acai berry is the fruit of the acai palm, which grows in parts of the Amazon rainforest. It tastes like a combination of forest berries and chocolate. The nutritional profile of this berry is a little different from those of 'regular' types of fruit, as it contains little sugar but many healthy fats. That's why this bowl fills you up more than the average smoothie.

The easiest way to include acai berries in dishes is in powdered form. Acai powder is brimming with antioxidants, essential vitamins, minerals, proteins, and healthy fats that support your metabolism and positively affect your cholesterol levels. Acai is often said to have anti-carcinogenic properties and to improve brain function. The fibre in the powder helps keep your gut in good working order.

Tip: The nutritional value of acai is complemented by the toppings you use. For example, opt for banana and/or granola with nuts and seeds, or, an equally excellent choice, bee pollen.

Pollen grains are an essential food to keep any bee colony strong and healthy, but humans can benefit from their healthy properties, too! Bee pollen is bursting with proteins, carbohydrates, healthy fats, minerals, and vitamins. It gives you power and energy, restores balance to the workings of your gut, and detoxifies your body. Bee pollen also gives you a helping hand if you're trying to lose weight.

INGREDIENTS (SERVES 2)
- 180 g (6 oz) strawberries (frozen)
- 150 g (5 oz) blueberries (frozen)
- 100 g (3½ oz) banana (frozen)
- 2 tbsp acai powder
- 2 tbsp honey (or more if you like it sweeter)
- 150 mL (5 fl oz) plant-based milk
- 1 tbsp peanut butter or other nut butter of your choice
- Toppings: bee pollen, chia seeds, coconut flakes, banana ... the choice is yours!

Tip: For this acai bowl, all the ingredients must be frozen to achieve the desired thick consistency, which makes it better for sprinkling toppings onto. The harder your ingredients are, the thicker your bowl will be. So don't take the fruit out of the freezer until just before you use it.

Tip: Peel and chop the banana before freezing: this makes it easier to whizz up in the blender.

METHOD
- Put all the ingredients in a blender and blitz them. Pour the smoothie in a bowl, and garnish with your toppings.

Tip: Add a little extra plant-based milk if your smoothie is too thick. If it's too thin, just add more frozen fruit.

Anouk's Goji Ketchup: The Healthy and Antioxidant-Rich Version!

Main player: goji berries
Other antioxidants: tomatoes (lycopene)

The secret ingredient in this ketchup is ... goji berries!
Sorry, what?
Yep, goji berries contain a big dose of antioxidants, but are also a good source of vitamin C. So, don't hesitate — add them!

INGREDIENTS (SERVES 6)
- 40 g (1½ oz) dried goji berries
- 20 g (1 oz) semi-sundried tomatoes
- 1 can of whole, peeled tomatoes (400 g / 14 oz)
- 1 tbsp cider vinegar
- 1½ tbsp honey
- ½ tsp garlic powder

- 1 tsp olive oil
- ½ tsp salt, a pinch of pepper
- 1½ tsp curry powder (optional)

METHOD

- Soak the goji berries and the sundried tomatoes in water for about 20 minutes. Pour off the water, and drain the juice from the tinned tomatoes.
- Mix all the ingredients in a food processor (or blender).

Tip: To turn it into curry ketchup, add 1½ tsp of curry powder.

Tip: In a well-sealed container, this ketchup keeps for several weeks in the fridge!

Rainbow Soup

This soup really deserves its rainbow name — as it contains the whole range of colours. After the rain(bow) comes the sun, and this soup is a similar little bit of happiness you can create for yourself. The antioxidants come from all the colours of the rainbow: bell peppers (green), carrots (orange), onion (red), cauliflower (purple), ginger (yellow), turmeric (more orange), and coconut (white).

INGREDIENTS (SERVES 2)

- 400 g (14 oz) carrots
- 1 red onion
- 1 green bell pepper (capsicum)
- ½ head of purple cauliflower
- 3 cm (1 in.) piece of fresh ginger

- 2 cloves of garlic
- A good glug of olive oil
- 1 can of coconut milk
- 1.2 L (1¼ qt) water
- 1 tbsp stock powder
- 3 tbsp soy sauce

METHOD

- Roughly chop the carrots, onion, and pepper. Cut the cauliflower into florets. Peel and finely chop the ginger.
- Fry the onion, the crushed garlic, and the ginger in a little olive oil in a pan. Add the carrots, bell pepper, and cauliflower, and cook together for a few minutes. Season with turmeric, curry powder, salt, and pepper.
- Add the coconut milk, water, powdered vegetable stock, and soy sauce, and leave the soup to simmer for 20 minutes on a medium heat. Liquidise with a hand blender.
- Add a little more water if the soup is too thick.

Matcha Nice-Cream

Main players: matcha powder, avocado, cacao nibs
Extra boosters: lime, (powdered) ginger, honey

Matcha is the powdered form of Japanese green tea. It's brimming with antioxidants (including EGCG) and fibre, and contains L-theanine (known for its calming effect and stress-reducing properties).

As well as containing a powerful green antioxidant, this ice cream also contains an unusual kind of fruit — avocado! Avocado is a source of antioxidants, omaga-3 fats, and vitamin E.

Honey is rich in antioxidants, thanks to the phenols in it, and it also contains enzymes, vitamins, minerals, and trace elements. Never heat honey to more than 40 °C (104 °F), as this destroys all its positive properties. Processed, industrially produced honey has none of those properties left anyway, so be sure to seek out the naturally produced kind (or buy it straight from your local beekeeper). You can certainly also sweeten this ice cream with agave or maple syrup if you are looking for a more budget-friendly option.

Topping: cacao nibs! A truly pure option, which also gives you an antioxidant boost! Cacao nibs contain an impressive number of nutrients: they are full to the brim with fibre, protein, and healthy fats. And they are far from lacking in minerals, too, providing magnesium, iron, zinc, copper, and phosphorus. Cacao nibs are the least-processed of all cocoa products, containing considerably less sugar, and are therefore a healthier alternative for all the chocoholics among us. ;)

This is truly a first-class dessert, so enjoy every mouthful! *YES! Dessert, pleazzze ... !*

INGREDIENTS (SERVES 4)

- 2 avocados
- 2 tbsp lime juice
- 300 mL (10 fl oz) coconut milk (canned)
- 1 tbsp vanilla extract
- 2 tsp matcha powder
- 2 tsp powdered ginger (or a quarter-to-half-inch piece of fresh ginger, grated)
- 50 g (2 oz) honey (you can adjust the quantity according to taste)

- A pinch of salt
- 1 ripe(!) banana
- Cacao nibs (optional)
- Mint leaves (optional)

METHOD

- Put all the ingredients (except the cacao nibs and the mint leaves) in a food processor or blender and mix until smooth.
- Pour the mixture into a freezer-proof container, and leave in the freezer to set overnight.
- Serve in a bowl, and garnish to taste with cacao nibs and a few mint leaves.

RECIPES TO SUPPORT YOUR MICROBIOME

Prebiotic Pasta Primavera

Prebiotic foods include leeks, chicory, onions, garlic, asparagus, and whole-wheat pasta.

Leeks and chicory contain inulin fibre, which supports the growth and activity of good bacteria in the gut. This substance can't be absorbed by your small intestine, so it's at the full disposal of the good bugs in your large intestine.

Onion, leeks, and asparagus contain prebiotic fructo-oligosaccharides, a type of roughage you can get from several kinds of bulb and stem vegetables, and which provides the best feeding ground in your gut.

As well as providing trace elements from more healthy grains, whole-wheat pasta is also a source of glucans, which you read about earlier in this book. They support the immune system's macrophages and help keep our defences up.

INGREDIENTS (SERVES 4)

- 2 cloves of garlic
- 1 large onion
- 2 leeks
- 200 g (7 oz) green beans
- 1 bundle of green asparagus
- 1 head of chicory
- 400 g (14 oz) whole-wheat pasta
- Olive oil
- 3 tbsp white wine
- 150 mL (5 fl oz) plant-based cream
- Salt and pepper to taste
- 1 tsp dried oregano
- 1 tsp dried thyme
- 1 handful of grated parmesan cheese
- Fresh parsley (optional)

METHOD

- Peel and finely chop the garlic. Slice the onion. Cut the leek into rings. Clean and halve the green beans. After removing the woody ends, roughly chop the asparagus. Remove the tough heart of the chicory, and cut into thin strips.
- Boil the pasta in salt water until done, as described in the instructions on the packet. Drain the pasta.
- Blanch the green beans and asparagus in salt water, drain, and put to one side.

- In a frying pan, fry the onion and garlic together in a little oil. Add the leeks and chicory. Fry together for a few more minutes.
- Remove the vegetables from the frying pan and put to one side. Pour the wine into the still-hot pan and let the alcohol evaporate. Add the green beans and asparagus, and fry in olive oil for a few minutes. Then add the pasta, the fried leeks and chicory, the plant-based cream, and the herbs and seasoning (pepper, salt, oregano, and thyme). Mix everything well.
- Serve the pasta in deep plates, garnishing with a little parmesan and, if you want, some parsley.

Green Goddess Smoothie

Prebiotics provide fibre and make for a fruitful feeding ground in the gut. The prebiotics in this smoothie come from the apple and the banana. The kefir provides probiotics, and even bacteria themselves. Also of nutritional interest in this smoothie are the lime juice (vitamin C) and honey (antioxidants). So, this apple, banana, and kefir smoothie is a vitamin and prebiotic bomb!

Just a word about kefir: it's rich in the bacteria *Lactobacillus acidophilus* and *Bifidobacterium bifidum*, contains a significant amount of lactic-acid bacteria, and is full of healthy yeasts. Traditionally, kefir is made by adding 'kefir grains' to water or milk and leaving the mixture to ferment. Its combination of beneficial microbiota makes kefir one of the most powerful probiotic foods. You can now find kefir for sale in most supermarkets.

INGREDIENTS

- 1 handful of spinach
- 1 small handful of mint (about 10 leaves)
- ½ a cucumber
- 1 apple, peeled
- 150 mL (5 fl oz) kefir
- 1 tbsp lime juice
- 1 to 2 tbsp honey (depending on how sweet you want your smoothie to be)
- 1 banana
- Fresh parsley (optional)

METHOD
- Put everything in a blender and mix!

Probiotic Vanilla-Cashew Yoghurt

INGREDIENTS

- 400 g (14 oz) unsalted, raw cashews
- 500 mL (17 fl oz) water
- 2 tsp probiotic supplement in powder or capsule form
- 2 tsp vanilla extract (optional)
- A pinch of salt
- 2 tbsp honey (optional)

YOU WILL ALSO NEED
- Preserving jars

METHOD
- Wash the preserving jars thoroughly with boiling water,

rinse them well, and dry them. Clean your work surface, use a clean chopping board and knife, and also wash your hands thoroughly before you begin!

- Soak the cashews in water overnight (make sure all the cashews are covered).

- Drain the cashews and rinse them thoroughly. Put the cashews, water, and probiotic powder in the mixer and blend to a smooth paste, about the consistency of mustard.

- Put the yoghurt in sterile(!) glass preserving jars, and cover with a clean towel. Don't fill the jars to the brim; make sure to leave some space at the top.

- Leave in a warm place to ferment for 24 to 36 hours.

- The yoghurt is ready when it becomes creamy in consistency and easily comes away from the side of the jar.

- When the fermentation process is finished, put the yoghurt in the blender again. Add the vanilla extract, a pinch of salt, and perhaps some honey. Mix again.

- Transfer to a clean preserving jar with a lid, and keep in the fridge.

Tip: This yoghurt will keep for up to a week in the fridge.

Fermented Carrots

You can adapt this recipe however you like, to include your favourite herbs and flavours, and change the amounts according to your preference.

Put these fermented carrots on your menu a few times a week to give your microbiome the best support possible. You can add them to a salad, for example, to make a great snack!

Vegetables are usually fermented using lactic-acid fermentation. The microorganisms present on the surface of the vegetables are mainly lactic-acid bacteria, or lactobacilli. They convert sugar into lactic acid, which makes your gut slightly more acidic and therefore a more hostile environment for pathogenic bacteria to thrive in.

INGREDIENTS
- Organic(!) carrots
- 2 cm (1 in.) piece of ginger
- 2 cloves of garlic
- ½ a red onion
- 50 g (2 oz) unrefined sea salt
- 1 L (1 qt) water

YOU WILL ALSO NEED
- Preserving jars with rubber seals

METHOD
- Be sure to use organically farmed carrots: pesticides have no place in the fermentation process. In addition, organic vegetables are more likely to carry the microorganism necessary for fermentation.
- Wash the preserving jar(s) very thoroughly in boiling water, rinse well, and dry them. Clean your work surface, use a clean chopping board and knife, and also wash your hands thoroughly before you begin!
- Wash the carrots thoroughly; there is no need to peel them. Cut the carrots into pieces — the size is entirely up

to you! I like to cut them into strips. Just make sure they
fit into your jar(s).

- Peel the ginger, garlic, and onion, and chop them finely.
- Put all the ingredients into the preserving jar(s).
- Make up a 5 per cent salt solution by dissolving 50 g (2 oz)
  of sea salt in 1 L (1 qt) of water.
- Fill the jar with the brine solution, making sure all
  the carrot pieces are fully covered by the brine. This is
  important during the fermentation process for two
  reasons:
  1. No oxygen must be allowed to reach the carrots, to
     make sure no unwanted bacteria can survive. Lactic-
     acid bacteria, by contrast, are anaerobic and can
     thrive in an oxygen-free environment.
  2. Very few bacteria can survive in a salt solution, but,
     luckily, lactic-acid bacteria love it!

Caution! Make sure the jars are not full to the brim with
salt solution: leave a couple of centimetres (about an inch)
free at the top. This is because fermentation releases gases
and will otherwise make the jar overflow.

- Fill a plastic freezer bag with the same salt solution, place
  it on top of the carrots, and close the jar. Don't use the
  rubber seal!
- Leave the carrots to ferment at room temperature for
  about two to three weeks. Try them at regular intervals
  (always using a clean fork!), releasing the built-up gas as you
  do so. Make sure that the carrots always stay completely
  covered by the brine. They are ready when they begin to

taste somewhat sour and no new gas is being released. If that hasn't happened by the end of your waiting time, just wait longer! You can stop the fermentation process after as few as three days, but the longer you leave them to ferment, the more flavourful they will become.

- When the carrots are completely fermented, close the jar with the rubber seal. Keep in the fridge to prevent the fermentation process from starting again.

- You can keep the carrots in their unopened jars for several months in a cool place. Once the jar is open, it will keep for about a month in the fridge.

- If the carrots start to taste strange or smell bad, throw them away. Something has also gone wrong if the liquid goes cloudy. Successful fermentation smells acidic.

SPECIALS

Immunity Shots

This powerful, detoxifying ginger-and-lemon combo is easy to whip up in a blender, and you don't need to have a slow juicer in your house to make it.

Research shows that taking a ginger-and-lemon shot every now and then can help with unexplained skin problems, support your muscles, promote digestion, and prevent nausea. As well as having anti-inflammatory powers, these shots are also a detox agent, cleansing your whole system and providing it with antioxidants.

I like to add a little coconut water for extra punch! Unsweetened coconut water is rich in vitamin C, electrolytes, potassium, and magnesium — just the minerals people are usually deficient in.

INGREDIENTS (SERVES 4)

- 3 lemons
- 15 g (½ oz) ginger
- 50 g (2 oz) honey
- 3 tbsp coconut water

METHOD

- Squeeze the lemons. Peel and chop the ginger.
- Mix all the ingredients in a blender. If you like ginger pulp, you can drink the shot as it is. Or you can sieve the mixture (this is how I prefer my shots).
- This little drink can be kept for several days in the fridge.

Cauliflower and Mushroom Risotto

We've already mentioned the health benefits of some kinds of fungi. Mushrooms are chock-full of glucans. Onion, garlic, parsley, and turmeric complement this immunity-supporting array.

INGREDIENTS (SERVES 2)

- 1 onion
- 250 g (9 oz) chestnut mushrooms
- 2 cloves of garlic, crushed
- Olive oil
- 400 g (14 oz) cauliflower rice
- 150 mL (5 fl oz) oat or soya cream
- 1 tsp turmeric

- Salt and pepper
- 70 g (2½ oz) grated parmesan cheese
- 1 handful parsley, finely chopped
- 1 handful chives, finely chopped

METHOD

- Chop the onion, wipe the mushrooms clean, and slice them. Fry the onion and garlic in a hot pan with olive oil. Add the mushrooms, and fry for about another 3 minutes.
- Add the cauliflower rice and the plant-based cream, turmeric, salt, and pepper, and cook uncovered until most of the liquid is gone and the cauliflower is cooked through.
- Turn off the heat, and stir the grated parmesan cheese into the risotto. Garnish with fresh parsley and chives.

The Immortal Immunity Burger

This is, without a doubt, the showstopper recipe in this book: unbelievably delicious, very simple to make, and brimming with immune-boosting nutrients!

Psyllium is ideal for upping your roughage intake. This dietary fibre can hold more than 40 times its own weight in water. Psyllium fibres have specific properties that stimulate gut function and bowel movements, as well as slowing down the body's absorption of carbohydrates and sugars. This helps prevent fluctuations in your blood-sugar level.

The other immortality effects of this burger come from turmeric, seeds, brazil nuts, salmon, lemon, and avocado, among others ... By now, you probably know yourself what nutrients each ingredient provides.

INGREDIENTS (SERVES 6)

- Boosters: turmeric, almond flour, pumpkin and sunflower seeds, salmon, brazil nuts, pine nuts, lemon, avocado, rocket (arugula) salad, red onion, garlic, lime, parsley, shiitake mushrooms

BUNS (MAKES 6)

- 180 g (6 oz) almond flour
- 7½ tbsp psyllium powder
- 1½ tsp salt
- 1½ tsp turmeric
- 1 tsp garlic powder
- 3 tsp baking powder
- 345 mL (11½ fl oz) water
- 3 tsp cider vinegar
- 4 egg whites (tip: keep the yolks — for homemade mayonnaise, for example!)
- 40 g (1½ oz) sunflower seeds
- 15 g (½ oz) pumpkin seeds
- 3 tbsp chia seeds

SALMON BURGER PATTIES (MAKES 6)

- 500 g (17 oz) fresh salmon
- 2 tbsp soy sauce
- 1 red onion
- 1 clove of garlic
- 1 tbsp mustard
- Juice of ½ a lime
- 1 handful of parsley

IMMUNITY BURGERS

- Brazil-nut pesto (see pp. 184–5)
- 250 g (9 oz) shiitake mushrooms
- 1 clove of garlic
- Olive oil
- Salt and pepper
- 150 g (5 oz) rocket (arugula) salad
- ½ red onion, cut into rings

METHOD FOR THE BUNS

- Preheat the oven to 180 °C (350 °F).
- Mix all the dry ingredients for the buns together in a large mixing bowl: the almond flour, psyllium powder, salt, turmeric, garlic powder, and baking powder.
- Bring the water to the boil.
- Add the cider vinegar and egg whites to the dry ingredients, and mix well. Pour in the boiling water, and mix the dough for about 30 seconds with a hand mixer. Take care not to overmix, otherwise the consistency of the dough will not be right!
- Mix the sunflower and pumpkin seeds into the dough.
- Cover a baking tray with baking paper. Wet your hands slightly, and roll the dough into six balls. Put the dough balls on the baking tray, and sprinkle them with chia seeds. Bake in the oven on a low shelf for 50 to 60 minutes. Tap the bottoms of the buns to test when they're done: if they sound hollow, the buns are ready.

Tip: You can make the buns ahead of time and freeze them. When you're ready to use them, simply warm them up in the oven.

METHOD FOR THE SALMON BURGERS
- Put all the ingredients in a blender and whizz them up.
- Roll the mixture into six balls with your hands, and then press them flat to form burger patties.
- Fry the burgers in a pan until golden brown.

FRIED SHIITAKE MUSHROOMS
- Cut the shiitake mushrooms into slices, and fry them in the pan with a crushed clove of garlic. Season to taste with salt and pepper.

ASSEMBLE YOUR IMMUNITY BURGERS
- Slice the buns through the middle, and spread with the brazil-nut pesto. Fill the bun with the salmon burger, fried shiitakes, rocket, and onion rings.

'Stay at home'

Dr Maggie De Blok

6

COVID-19

I've already referred to it in passing here and there, but of course I can't write a book about the immune system without going into more detail about the infamous coronavirus that holds the entire world in its grip as I write.

Coronaviruses refers collectively to a group of human respiratory viruses. They get their name from the Latin word for 'crown', or 'wreath', due to their appearance under a microscope. Coronaviruses are common in nature, and most cause little more than an ordinary cold, with a runny nose and a sore throat.

Every now and then, a virus like this mutates in such a way that the symptoms it causes become more serious. In 2002, the SARS — severe acute respiratory syndrome — virus mutated in the body of a bat or a civet cat, and then mutated again to make it transferable to humans, where it turned out to be highly pathogenic. The MERS — Middle East respiratory syndrome — virus jumped the species barrier from bats to camels and then to humans. The novel (new) coronavirus that causes COVID-19 followed a similar path of mutations, this time most likely from bats to pangolins to humans.

Facts and Figures

We're currently being bombarded from all sides with stories, facts, and pseudo-facts, so that we can barely see the wood for the trees. It's normal to be afraid in uncertain times, but that's precisely when we need to focus on actual facts. And we have to count on the epidemiologists and virologists among our scientists to provide us with those facts.

Case-Fatality Ratio

The case-fatality ratio is one of the parameters we look at to see how serious a disease is. It's the number of patient deaths from the disease, divided by the number of patients infected with it, multiplied by 100. The SARS virus had a case-fatality ratio of 10 per cent, while the figure for MERS was far more serious, at 34 per cent. The case-fatality ratio for COVID-19 is still difficult to calculate. The number of confirmed cases varies a lot from country to country, which is due to the different ways patients are tested for the disease, and the number of tests carried out. There's also some debate about the absolute number of deaths from COVID-19, as the cause of death is reported differently in different countries. So, we must always be very cautious when dealing with the available figures regarding this specific virus, and especially so when we try to interpret them.

R Number

Another epidemiological parameter we can look at is the R number, or basic reproduction number, which is an expression of how easily a disease spreads. The basic reproduction rate of the initial strains of COVID-19 is in the region of two to three: each infected person goes on to infect between two and three others, statistically speaking. Those three newly infected people in turn go on to infect two or three more. Those nine people go on to infect three more each, and, before we know it, 27 people are infected with the coronavirus. This shows that the virus that causes COVID-19 is quite infectious — easily transmitted from one person to another. For comparison, a classic flu virus has an R number of 1.3. That means it is less infectious, and the increase in the number of people infected is much smaller.

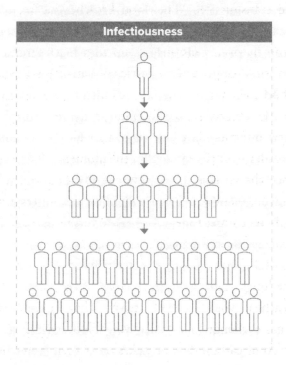

Infectiousness

Transmission

COVID-19 is a respiratory infection, so it mainly affects the respiratory tract. There are indications that it can spread via faecal–oral transmission. You probably don't ever mean to put your faeces in your mouth, but make no mistake: it happens more often than you'd like to think. Perhaps you have sometimes forgotten to wash your hands after going to the toilet ... ? For this reason, it's important to make sure that toilets used by infected people are thoroughly disinfected, and that general hygienic rules are followed under all circumstances.

However, the main way the virus spreads is through particle transmission. When you breathe, sneeze, cough, or speak, you spread tiny aerosols and airborne droplets. They can fly over a distance of up to a metre and a half. Transmission happens when particles of an infected person's saliva find their way to the mucous membranes of another person. No symptoms usually appear in the first four to five days after infection. But beware, you are already a carrier — and spreader — of the virus from the moment of infection!

Once the virus is in your system, it enters the cells of your mucous membranes, especially those in your respiratory tract, where it starts to reproduce. You've already read about that process in some detail earlier in this book.

COVID-19 has a distinct preference for burrowing deep into the lungs, right down to the level of the alveoli. When you breathe in, your alveoli fill with air. They are the location where gas exchange takes place: your blood enters, releases its burden of carbon dioxide, takes on oxygen from the air, and leaves on its journey around the body. If the virus gets into the alveoli, the body will activate its inflammatory response, resulting in the alveoli being filled with fluid and pus — which is one of the body's immune responses. This hinders the process of gas exchange, as there's less space for air in the lungs, leading to oxygen deficiency.

The Tests

How can you tell whether your gas exchange system is still working properly? There's only one way to find out: using a device to measure your blood's oxygen-saturation level. Normal, healthy people should have a blood-oxygen-saturation level of between 95 and 100 per cent. A level of 90 to 95 per cent can already cause problems for some people. If saturation dips below 90 per cent, the body's organs can't get enough oxygen to keep functioning properly. At that point, breathing needs to be assisted so that all parts of the body can get the oxygen they need.

Of course, a low saturation level doesn't necessarily mean you're infected. The only way to find that out is to test whether you are carrying the virus. The PCR test — the classic swab test — in which a sample is taken from the mucous membranes in your nose or throat, or both, and then tested in the laboratory for traces of the coronavirus's RNA. This test answers the question of whether the virus is present in your body or not.

A different test can tell you whether you're already immune to the disease, by examining your blood for antibodies. You've already read in previous chapters of this book that the presence of antibodies in your system usually means you have developed an immunity to a given disease.

That said, we are, however, still not 100 per cent sure whether the mere presence of antibodies is equivalent to immunity in the specific case of COVID-19. We *can* be sure whether you've been in contact with the disease and whether your system has produced antibodies. This is actually a measurement of the status of your immune system: if there are lots of antibodies present, it's a clear sign that your immune system is fighting or has fought something.

We now also know that the coronavirus causes some people's immune systems to overreact or to continue fighting for a long time. Their cells produce large amounts of signal substances (cytokines), whipping up what's called a cytokine storm. This means some people get very sick, the normal functioning of the alveoli in their lungs is reduced over a very long period of time, and they become vulnerable to secondary infections. Lungs that aren't functioning properly are always more susceptible to bacterial attack than normally functioning organs.

What Can You Do Yourself?

There are three main ways you can help yourself:

- Follow the government's regulations. The people working both in the public eye and behind the scenes know their business — it doesn't always look that way, I know, but believe me: the guidelines are science-based and are meant to protect you, not oppress you. And, by following them, you're also protecting others! One of the guidelines covers hand-cleaning, both by washing with soap and water, and by using hand sanitiser. The only correct way to wash your hands is the technique described in detail earlier in this book, which happens to be the technique used by surgeons. Social distancing and the wearing of mouth-and-nose coverings are also essential ways of protecting yourself and others. When you need to cough or sneeze, do so into the crook of your elbow. Of course, make sure you use the other arm to 'elbow bump' people when you greet them — shaking hands is out of the question. Try to avoid contact with surfaces where the virus may be hiding. The virus can survive for about three hours in the air, but up to 24 hours on cardboard or paper surfaces, and it can easily survive for up to 72 hours on plastic or stainless steel. Regular cleaning is the message.

- Improve your lifestyle as much as you can. Half this book is dedicated to this point and so I don't have much more to add here. Suffice it to say: the healthier your lifestyle is, the better your immune system will work. Exercise, nutrition, and sleep are super-important, but the health impact of stress is also a factor that shouldn't be underestimated, especially in these uncertain times. When the causes of stress are too much or persist for too long, chronic stress and a disrupted immune system can result. Lack of certainty and loss of control can drag anyone into a downward spiral. 'Keep calm and carry on' may be a massive cliché, but, in this case, it's the best advice I can give you. And take regular breaks from the news and social media!

- As you follow the general regulations, you can also take action on a personal level by talking to your doctor or another professional. Measure to manage: find out how you can tweak your own behaviour for more security, and ask what aspects it makes sense for you, specifically, to re-examine. A large number of people are high-risk patients: those for whom COVID-19 can have more dangerous consequences. These include people with cardiovascular disease, diabetes, obesity, lung disease, cancer, and certain auto-immune conditions.

- What deficiencies show up in your blood tests, and how can they be tackled? I'm thinking mainly of vitamin D here, a hormone that has an important influence on your white-blood-cell count, as well as on your cardiac health and the healthy functioning of your nervous system. Cortisol, the stress hormone, is also an important parameter. A stressful environment — which is pretty much every human environment right now — causes cortisol to peak throughout the day, making it more difficult for levels to come back down in the evening. This interferes with your sleep, which in turn undermines your immune system. These are just two examples — I have covered many more in greater detail in the 'Lifestyle' chapter of this book.

To Boost or Not to Boost

I already asked the question in the Introduction: can you 'boost' your immune system and improve its functioning over the short term? No, I don't believe you can. But when it comes to nutrition and dietary supplements, I do believe you can give your immune system some extra support. That's why we need to start at the bottom of the pyramid and make sure all elements are in equilibrium.

Water really is the stuff of life — and it's one of the purest gifts you can give your body. Followed by a plant-based diet, which should include all the colours of the rainbow. Of course, you also need all the necessary building materials, which you can get in the form of protein-rich fish, meat, or vegetarian alternatives. You may then want to add supplements to your diet — the opposite approach, starting with the supplements and planning your diet around them, makes no sense, as I see it. The taking of dietary supplements is best managed in consultation with your doctor. Discuss your lifestyle with your doctor — on the basis of test results or not — and together work out what you need and what will benefit you the most. And remember: exercise, sleep, rest, live!

Stay healthy, everyone, and take care of yourselves!

ACKNOWLEDGEMENTS

Words of thanks ...

... are the last thing you write, after you've taken it upon yourself to write a book.

And that's when you think *Uff! Finished at last!* Combined with a feeling of letting go. After all, when the final, proofed version is in my mailbox, I can't make any more changes to it. So, all I can do is wait and hope that this new work will inspire you and encourage you to take more care of yourselves and others.

It only remains for me to thank everyone who made this book possible. It would be impossible to mention all of them by name, so what follows is a necessarily incomplete list.

Marie D'haenens, for her editing work, and for asking the questions you would ask as readers, allowing me to anticipate them and work the answers into the text.

Anouk Brackenier — thanks over and over and over again, both for the recipes and for your passion for health.

Steven, Kristof, Pieter, Julie, Kim, and everyone else in the publishing team, for believing in me and letting me do my thing. And, above all, for letting me flirt a little with the deadline — it made it easier to write about how too much stress can be bad for your health, and made me realise that I'm probably the biggest source of stress to myself.

Lien, Jonas, Paul, Carolle, Tessa, Sander, Evelyn, Lennart, Bavo, Els, Marie, Jelle, Steven, Thomas, Thijs, Dieter, Stefan, Stein, Michael, Kristof, Sammy, Barbara, Chris, Ken, Kris, Maarten, Pieter, Jasper, and all the other experts in their field who I had the privilege of working, brainstorming, creating, and carrying out ideas with, and who challenged me every day to give the best of myself.

And, finally, *you!*

To write something that never gets read is pretty senseless, so thank you all for reading this.

Thank you for taking away from it whatever works for you.

Thank you for sharing that with others and with me.

Thank you for providing the quotes for my chapters ...

I hope to meet you all in a happy, healthy, and free future!

Stay Safe & Take Control of Your Health!

Servaas
@servaasbinge